TOMORROW'S WAR, TODAY'S DECISIONS

IRAQI WEAPONS OF MASS DESTRUCTION AND THE IMPLICATIONS OF WMD-ARMED ADVERSARIES FOR FUTURE U.S. MILITARY STRATEGY

by Robert W. Chandler

with Ronald J. Trees

AMCODA Press

McLean, Virginia

Copyright 1996 by Robert W. Chandler

ISBN   0-9650770-0-4

Library of Congress Catalog Card Number: 96-84124

Published by AMCODA Press, McLean, Virginia 22101

Printed in the United States of America

In Memoriam:

Brigadier General Garyl "Sip" Sipple, U.S. Air Force
Husband, Father, Warrior, Friend

What I fear is not the enemy's strategy,

but our own mistakes.

— *Pericles to the Athenians*
on the Spartan ultimatum,
432 B.C.

Table of Contents

LIST OF FIGURES ix

LIST OF TABLES x

PREFACE xiii

EXECUTIVE SUMMARY xv

1- PROLIFERATION AND THE FUTURE OF INTERNATIONAL
    CONFLICT 1
    The Proliferation of Knowledge and Technology . . . 2
    Iraq's Weapons of Mass Destruction 11
        The Persian Gulf War
        Post-War Discoveries by U.N. Inspectors

2 - BUILDING A NUCLEAR ARSENAL 21
    The Persian Gulf War 22
    Iraq's Nuclear Weapons Program 27
        Petrochemical Project 3 (PC 3)
        IAEA On-Site Inspections
        Uranium Mining, Production, and Processing Sites
        Sites Related to Uranium Enrichment
        Sites Related to Weaponization
        *Other Sites Supporting Iraq's Nuclear Weapons
            Program*
        The Iraqi Crash Program
    Implications for Counterproliferation Operations . . . 57

3 - **SHADING A BIOLOGICAL WEAPONS PROGRAM** 63
    The Persian Gulf War 66
    Iraq's Biological Weapons Program 70
        U.N. Inspections
        A Locked Chicken House
    Implications for Counterproliferation Operations . . . 79

4 - **EXPANDING A CHEMICAL WEAPONS PROGRAM** 83
    The Persian Gulf War 86
    Iraq's Chemical Warfare Program 90
    Implications for Counterproliferation Operations . . . 99

5 - **DEVELOPING A BALLISTIC MISSILE PRODUCTION**
    BASE 103
    The Persian Gulf War 107
    Iraq's Ballistic Missile Program 114
    Implications for Counterproliferation Operations . . . 125

6 - **DECEIVING WESTERN EXPORT CONTROLS** 129
    International Export Controls 130
        Nuclear Export Control Regimes
        Biological and Chemical Export Control Regimes
        Ballistic Missile Export Control Regime
    Iraq's Foreign Procurement Programs 137
    Implications for Counterproliferation Operations . . . 143

7 - THE IMPLICATIONS OF REGIONAL WMD
PROLIFERATION FOR U.S. MILITARY STRATEGY:
ASSESSMENT 147
   Lessons Learned From the Iraqi Experience 147
      *Regional WMD Proliferation Is No Longer
      Tomorrow's Problem*
      *Regional WMD Proliferation Is Likely To Continue
      Unabated*
      *Monitoring Regional WMD Proliferation Will Be
      Extraordinarily Difficult, Particularly During
      Crises And Conflict*
      *As a Distinct Target Set, WMD Programs Present a
      Daunting Military Challenge*
   Implications of WMD-Armed Adversaries on the
      U.S. Military Posture 156
      Deterring U.S. Intervention
      Intimidating America's Regional Allies
      *Disrupting and Blocking U.S. Deployments and/or
      Combat Operations*
      Limiting U.S. War Aims
   Conclusions: A Snapshot of the Future 162

8 - THE IMPLICATIONS OF REGIONAL WMD
PROLIFERATION FOR U.S. MILITARY STRATEGY:
OPTIONS 165
   Clinton Administration Policy: The
      Counterproliferation Initiative 165
   Assessing U.S. Options for Countering
      WMD-Armed Regional Adversaries 171
      Nuclear Deterrence
      Theater Defense
      Counterforce Operations
   Conclusions: The Road Ahead 198

SELECTED BIBLIOGRAPHY 205

INDEX 219

Figures

2-1 Eight Nuclear Targets Bombed During the
     Gulf War 23
2-2 Al Tuwaitha 31
2-3 Nineteen Primary Sites Related to the Iraqi
     Nuclear Weapons Program 37
2-4 Al Jesira 40
2-5 Tarmiya 46
2-6 Al Atheer 52
2-7 Fifty-Six Iraqi Nuclear Facilities 59

3-1 Seven Iraqi Biological Weapons Facilities 75

4-1 Twenty-Three Iraqi Chemical Facilities 98

5-1 Twenty-Nine Iraqi Scud Facilities 123

Tables

1-1   A Three-Tiered International System   3
1-2   Suspected Weapons of Mass Destruction
         Programs   8
1-3   Technical Hurdles for Nuclear, Biological,
         and Chemical Weapon Programs   12
1-4   Variation in Number of WMD and
         WMD-Related Installations   17

2-1   Eight Known and Suspected Nuclear Facilities
         During the Gulf War   24
2-2   Al Tuwaitha Buildings Involved in
         Weaponization and Enrichment   32
2-3   Iraq's Ten Core Nuclear Facilities   36
2-4   Uranium Mining, Production, and Processing
         Sites   38
2-5   Sites Related to Uranium Enrichment   42
2-6   Sites Related to Weaponization   50
2-7   Other Sites Supporting Iraq's Nuclear
         Weapons Program   54

3-1   Some Common Biological and Toxin Agents   67
3-2   Iraq's Biological Warfare Facilities   76
3-3   Biological Monitoring Sites   79

4-1   Chemical Facilities   94
4-2   Chemical Warfare Items Destroyed at
         Al Muthanna 100

5-1 Known Iraqi Ballistic Missile Targets 113
5-2 Iraq's Ballistic Missile Facilities 120

6-1 Sample Foreign Sources of Iraqi Supplies 138
6-2 Circumvention of Biological, Chemical,
     and Missile Export Controls 142

Preface

This study is about America's current military strategy for protecting vital interests overseas from regional aggression. A composite sketch of a 21st century proliferator is drawn from Saddam Hussein's extraordinary nuclear, biological, and chemical weapons and ballistic missile programs that turned out to be more elusive and far greater in size and complexity during the Persian Gulf War than was imagined by the U.S. intelligence community. Several U.S. military strategy options are assessed. Emphasis is placed on how today's strategy decisions will influence the force posture that will shape America's response to tomorrow's war.

Three principal information sources are used in this case study. First, the U.S. Air Force "Gulf War Air Power Survey" is drawn upon for details of the nuclear-biological-chemical-missile targets that guided the Coalition's 1991 air operations. Secondly, the post-war on-site inspection reports by the International Atomic Energy Agency and the United Nations Special Commission on Iraq provide a great deal of insight to Baghdad's programs for weapons of mass destruction and missiles. These reports suggest a far more elaborate nuclear, biological, and chemical weapons and missile target base than was prosecuted in the Gulf War. Finally, selected secondary sources were especially useful, in particular Michael Gordon and Bernard Trainor's *The Generals' War: The Inside Story of the Conflict in the Gulf.*

Ronald J. Trees provided much of the structure for this book in his pre-study on the Iraqi weapons of mass destruction and their related facilities. His keen insights to the problems encountered by Coalition air planners in targeting Iraq's special weapons has been especially useful. Special thanks also go to Trisha Scianna who maintained "command" over the manuscript preparation process and prepared many of the graphics.

Robert W. Chandler

Executive Summary

As America heads into the 21st century, U.S. national security strategy is focused on protecting vital interests abroad from regional aggression. Regional aggressors are now expected to be armed with nuclear, chemical and biological weapons—the so-called "weapons of mass destruction" (WMD)—and the means to deliver them. How best to deal with WMD-armed adversaries has emerged as the subject of an intensifying debate in the national security and political communities.

This book suggests that America's current approach for countering the wide-ranging regional WMD threat is inadequate and that a bold shift in direction is needed if the United States wishes to protect its international interests in future. As highlighted by this analysis, proliferation of WMD represents a fundamental change in the international security environment that cannot be addressed by "tinkering at the margins." To that end, we recommend that the United States exploit its technological edge to develop a force posture and military strategy that emphasizes theater missile defense and long-range strike systems capable of deterring and defeating regional aggression from beyond the range of enemy weapons of mass destruction.

Lessons Learned From the Iraqi Experience

Our analysis first details Iraq's drive to field WMD and U.S.-led efforts to neutralize Iraqi WMD capabilities during Desert Storm. The value of this comprehensive review is in the keen insight it provides for future planning. The lessons we have learned from the Iraqi experience yield a composite sketch of the 21st century proliferator. With this in hand, U.S. decisionmakers and military planners can now make legitimate assumptions about the regional WMD threat and begin constructing the most effective approach for countering it. Four main conclusions can be drawn from our analysis of Iraq's ambitious program for developing WMD and the Coalition campaign to neutralize it during the Gulf War:

- **Regional WMD proliferation is no longer tomorrow's problem.** Before Desert Storm, U.S. policymakers generally believed that regional WMD proliferation could be forestalled. The stunning series of post-Gulf War revelations about the size, scope, sophistication and maturity of Iraq's nuclear, chemical and biological weapons and ballistic missile programs captured in this report have changed all that. Washington now acknowledges that regional WMD proliferation is upon us and that U.S. strategy must now be reoriented to deal with it.

- **Regional WMD proliferation is likely to continue unabated.** Regional states have powerful incentives to acquire WMD. Rogue states in search of regional hegemony seek WMD to intimidate and/or defeat their neighbors and to deter and, if necessary, disrupt U.S. intervention. Meanwhile, as more and more states proliferate, even non-aggressor states will feel compelled to follow suit to deter WMD-armed aggression. The Iraqi case also shows that proliferation is all but impossible to prevent, that Western nonproliferation efforts have little effect on a determined proliferator. Finally, many fear that nuclear expertise, advanced nuclear technology, high-grade nuclear materials, and possibly even full-up nuclear weapons will be available to the highest bidder from the cash-strapped former Soviet republics.

- **Monitoring WMD proliferation is extraordinarily difficult, particularly during crises and conflict.** In retrospect, pre-Gulf War WMD-related intelligence was woefully inadequate. The Coalition air campaign planners, for example, targeted just eight Iraqi nuclear facilities during the war; in contrast, International Atomic Energy Agency inspectors discovered fifty-six such sites afterward. Before Desert Storm, experts pegged Iraq as several years from fielding even a crude nuclear device; it is now believed they were just twelve to eighteen *months* away from producing one or more nuclear devices and/or deliverable weapons. It would be wrong, however, to cast full blame on the intelligence community. Iraq went to extraordinary lengths to shroud the development of its WMD triad; we can expect future proliferators to follow suit.

- **As a distinct target set, WMD programs present a daunting military challenge.** Using the Iraqi WMD program (January 1991) as a baseline target set for future counterforce (counter-WMD) operations, one is immediately struck by the sheer scale of the projected effort. Planners would have to target over 240 "fixed" facilities and hundreds of other dispersed and/or mobile assets. Moreover, such an operation would be complex, technologically challenging, politically controversial, and very risky. The proliferator would undoubtedly know America viewed his WMD development as a burgeoning threat and that military action was a possibility. Defensive preparations, such as concealing and dispersing critical WMD assets and placing air defenses in a higher state of readiness, would likely be stepped up. Facilities could be located in hardened and/or deeply-buried bunkers resistant to all but the most advanced penetrating weapons and virtually invulnerable to current-generation cruise missiles; guarded by sophisticated air defenses; and located in, or indeed relocated to, heavily-populated urban areas.

The Implications of Regional WMD Proliferation for U.S. Military Strategy: Assessment

Potential U.S. regional adversaries seek WMD not only to intimidate and/or defeat their neighbors, but to deter and, if necessary, disrupt and block outside intervention. Most, if not all, such states view the United States as their principal extra-regional threat, and WMD as the means for "devaluing" U.S. military might by exploiting America's aversion to casualties and its clear dependence upon access to ports, airfields and military facilities in the theater of conflict. Adversaries may have several different objectives for utilizing WMD against the United States:

- **Deterring U.S. Intervention**. An aggressor might seek to deter U.S. intervention by convincing the U.S.

leadership that the expected costs (i.e., casualties) outweigh the possible gains. Threatening WMD against U.S. assets enroute to, or already in, the theater of conflict is one obvious strategy. American troops, ports, airfields, bases and naval forces in the region would form attractive targets. It is important to note that an aggressor does not need to actually use, or even threaten to use, his weapons to achieve this objective. As long as the United States is reasonably certain the enemy does in fact possess deliverable WMD, the potential for mass casualties could very well cause the president to balk at sending American forces abroad, or to backpedal after a U.S. deployment has already begun.

- **Intimidating America's Regional Allies.** U.S. military intervention in a regional crisis could hinge on the formation of an international coalition with allies in the theater. Moreover, the success of America's current regional warfighting strategy (deploying large numbers of forces into the region) depends upon access to bases in the theater. A regional aggressor might threaten WMD use to prevent the formation of such a coalition, splinter an existing coalition, or coerce neighbors into denying the United States access to their ports, airfields, military facilities, and airspace. Lack of coalition support would increase domestic opposition to the use of American force abroad and could delay a U.S. military response; denial of theater access would seriously circumscribe U.S. power projection capabilities.

- **Disrupting and Blocking U.S. Deployments and/or Combat Operations.** A WMD-armed aggressor could also employ his weapons to cripple on-going U.S. power projection operations. For example, the aggressor could launch chemical or biological strikes against such targets as main U.S. ports and airfields of debarkation to disrupt the flow of combat aircraft, troops, heavy military hardware, munitions, and other supplies into theater. If launched at the front of a short-warning invasion, such a WMD attack could buy an aggressor additional time to achieve his initial military objectives (e.g., occupation of the victim state's capital city or critical economic assets), dig in and disperse, and force the U.S. to decide between conceding or fighting a much more costly war of eviction.

- **Limiting U.S. War Aims.** Adversaries could also use or threaten the use of WMD to protect their regimes or states from total defeat by raising the perceived costs of defeating him beyond the expected gains. This logic applies evenly to nuclear, chemical and biological weapons precisely because all are capable of causing "mass destruction." Some analysts argue that when facing defeat, regional leaders may become "non-deterrable." The primary concern for many developing world regimes is retaining a hold on power. Faced with losing their power, regional leaders essentially have nothing to lose; in their minds, a WMD attack may be viewed as the last hope for survival.

The Implications of Regional WMD Proliferation for U.S. Military Strategy: Options

Current U.S. options for countering the regional WMD threat fail to mitigate these potentially debilitating problems. The current U.S. strategy for countering WMD-armed regional adversaries has three pillars: nuclear deterrence, defense (both active and passive), and counterforce. In reviewing each, the following points summarize our observations:

- **Nuclear Deterrence.** Deterrence is highly problematic in a regional context. For one, the credibility of the American nuclear threat is clearly limited. Unless intrinsic U.S. interests are present in the region, American nuclear threats may not be credible. Leaders with nothing to lose have little reason to fear nuclear retaliation, while those pondering chemical and/or biological strikes have good reason to believe such actions would not provoke a nuclear response. Perhaps most importantly, the concept of deterrence is inherently flawed as a policy for dealing with weapons that need not actually be employed by the aggressor to achieve warfighting objectives. The specter of casualties will linger over U.S. decisionmakers from day one in crises involving WMD-armed adversaries and could very well cause U.S. and allied leaders to balk at taking decisive action.

- **Active and passive defenses.** Traditional air defense systems will likely remain effective against WMD-armed strike aircraft and, to a somewhat lesser extent, cruise missiles for the foreseeable future. Due largely to budget cuts to cover "other priorities," that same optimism does not hold for theater ballistic missile defenses. Three "core" TMD systems are in development, but none have been seriously tested and all are subject to budget-related deployment slippages. Meanwhile, the threat is getting more complex. Ballistic missile warhead accuracies using the Global Positioning System will intensify threats to U.S. forces and America's regional friends and allies. Potential adversaries are expected to develop TMD countermeasures such as decoys and early-release chemical and biological submunitions. In short, the U.S. TMD forecast remains clouded with uncertainty. Passive defense measures such as force dispersal, detection and identification, individual and collective protection, decontamination, and casualty management can only be viewed in terms of damage control.

- **Counterforce.** Counterforce operations would be designed to "neutralize" enemy WMD before it can be used against U.S. forces and/or interests abroad. Against the most threatening WMD programs, however, current and planned U.S. counterforce capabilities are not up to the task. The primary problem here is America's heavy reliance upon theater-range aircraft for precision strike. To achieve the initial decisiveness critical to any counterforce operation, the U.S. would need to deploy numerous tactical aircraft within range of the target country. This sizable theater presence would, however, eliminate all elements of surprise and leave most of the American aircraft vulnerable to the very weapons they were sent to destroy. At the same time, preemptive counterforce operations, even if militarily feasible, would probably be politically impractical, both domestically and internationally.

Conclusions: The Road Ahead

Under the Clinton Administration's Counterproliferation Initiative, WMD-armed regional aggressors can limit U.S. power projection capabilities. America's three-pronged approach for countering the wide-ranging regional WMD threat is clearly inadequate. Deterrence can never be guaranteed; theater missile defenses have gaping holes in them as a result of budgetary restructuring to cover "other Department priorities;" and passive defense programs are underfunded. This leaves counterforce operations, but they will remain militarily infeasible so long as theater-based aircraft are left to carry the precision strike burden. There is, in fact, little the United States can do to negate the many implications of proliferation short of "extending" the theater of operations beyond the range of enemy WMD.

What alternative policy, strategy, and force structure options are available to U.S. decisionmakers and military planners? The United States is clearly capable of developing a robust, if imperfect, 21st-century counterforce capability. Our answer is simple and straightforward: By heavily augmenting its long-range precision striking power, particularly the heavy bomber and cruise missile forces, the United States can credibly threaten to exact a heavy toll on aggressor WMD programs worldwide without subjecting itself to the litany of problems faced by a largely theater-based force. Operational effectiveness will continue to be limited to some degree by various technological challenges, such as those associated with intelligence-gathering and hard-target penetration. However, "effectiveness" here is a relative concept. In the eyes of a potential proliferator, the damage maximization potential of a beefed-up and largely invulnerable long-range strike force could very well be enough to discourage WMD development and/or acquisition. In conflicts with known proliferators, such a capability would allow the U.S. conduct relatively risk-free counterforce strikes prior to large-scale, and otherwise highly vulnerable, theater force deployments. Absent the immediate threat of theater WMD retaliation, long-range counterforce operations could be protracted, allowing the U.S. to sustain strikes until it is deemed "safe" to enter the theater.

But this begs a larger, more fundamental, and far more important question: If the United States has within its technological and fiscal grasp the capability to strike this decisively from outside the range of WMD, why doesn't it move its entire military posture in this direction? Why should the core military strategy still entail sending American forces relatively unprotected into a proliferator's backyard, where the success of the strategy hinges upon our capability to deter, defend against, or defeat his WMD, none of which can be guaranteed?

The real question is not how can the United States deter WMD use, neutralize enemy WMD capabilities, or defend against WMD attack. The real question is how can the United States deter the regional aggression WMD will be used to support? The answer: by *demonstrating* the capability to stop that aggression with forces operating from beyond WMD range—forces invulnerable to theater WMD. Should deterrence fail, the U.S.

would then feel confident that it can take swift and decisive military measures to protect its threatened interests.

This is not to suggest that the United States should simply scrap its tactical, or short-range components, or even radically alter its force structure. Far from it. We are, however, recommending that the *burden* be shifted from forces dependent upon regional basing to those that can operate effectively from outside the region of conflict. The first phase of any U.S. attack or counterattack, and certainly any preemptive attack, is likely to be an air campaign. Air-delivered weapons can be used against strategic targets and/or advancing armies. Why should such a campaign not be staged from beyond the range of the weapons that can really hurt the U.S. expeditionary forces or cause American political and military leaders to balk at taking decisive action in the first place?

Long-range precision strike, even against mobile systems such as advancing armor and mobile missile launchers, is now technologically possible. Stealth bombers, capable of fighting with minimal support, can now accomplish tasks once reserved exclusively for fighter aircraft or fighter-escorted bombers. Cruise missiles, though less flexible than penetrating aircraft, offer many important advantages and will increase in capability through technological advances. Neither are vulnerable to theater-range WMD. The B-2 is clearly the ultimate manned strike platform for a WMD world both in a counter-force and general warfighting sense but the Pentagon argues we should spend future resources on new fighters (the F-22, F-18E/F, and Joint Strike Fighter), not bombers. The Air Force does not have a long-range cruise missile in production or development. Simply put, the United States is passing on what is clearly the best defense against weapons of mass destruction: being somewhere else when they go off.

As long as the United States continues to subscribe to the current strategy of pouring sizable numbers of American forces into a proliferator's backyard, any and all counterproliferation measures will be tantamount to "tinkering at the margins." This can lead to only one conclusion: the strategy is flawed. If the "plan" does not allow the United States to exploit its strengths and field the most appropriate fighting force possible, the plan is wrong and a new blueprint is in order.

In his May 1994 *Report on Nonproliferation and Counterproliferation Activities and Programs* to Congress, Deputy Defense Secretary John Deutch said that it was "not easy to change the direction of the ship of the state especially when its course for over 45 years was primarily aimed at preparing for threats that have receded, while the problems of proliferation have grown and become more urgent" (p. 25). It is, therefore, hardly surprising that the regional WMD threat is being considered in a vacuum, separate from mainstream military planning. Hamstrung by a budget-driven strategy-making process, the U.S. military establishment has had an exceedingly difficult time seeing the big picture as it struggles to adapt to the post-Cold War era. Yet, this cannot continue if the United States is to effectively counter the regional WMD threat. Military strategy must be designed to satisfy U.S. national security objectives.

The proliferation of weapons of mass destruction must be recognized as a fundamental change in the security environment, and the implications of WMD-armed adversaries must be accounted for at every stage of the military planning process. Unfortunately, the Pentagon, despite recent progress, still seems determined to plan for the wrong war and the wrong threat. The potential outcome in a future regional conflict could well be disastrous.

Chapter 1
Proliferation and the Future of International Conflict

The end of the Cold War provided U.S. decisionmakers with a classic challenge—adapting to the loss of a longstanding, central organizing principle. The motivating premises and assumptions of the Cold War had defined U.S. national purpose for more than thirty years. They dictated the thrust of our global security policy and compelled the acquisition of an unparalleled array of military capabilities. With the collapse of this guiding construct during 1991 and 1992, the United States was left to devise a new set of organizing principles around which to form a new consensus to shape the U.S. role in the world and to define the size, composition, and equipage of the armed forces.

The Persian Gulf War was a wake-up call. Thanks to the United Nations ongoing inspections of Iraq's weapons-making complex following the 1990–91 Gulf War, we now can make a detailed account of the vast scale of Saddam Hussein's nuclear, biological, chemical, and ballistic missile programs. As the world's first compleat proliferator of weapons of mass destruction (WMD), Iraq offers American policymakers and military planners a rare glimpse into an unfolding future filled with deadly new threats and grave risks as the up-gunning of the Third World continues worldwide.

The global proliferation of nuclear, biological, and chemical weapons—and the missiles, aircraft, and imagery and navigation satellites necessary to deliver them accurately across great distances—profoundly changes the stakes in U.S. security interests, policies, and strategy. The stakes can include the lives of thousands of people who may become targets or, in the extreme, entire countries could be placed at risk. These kinds of stakes will influence heavily the U.S. propensity to act in the face of threats to regional allies and friends and, having decided to act, on the particular actions the U.S. will be willing to take. Eventually, when the United States itself is placed at risk by the proliferation of nuclear, biological, and chemical weapons, the stakes will be increased even more, bringing yet another round of shifting U.S. interests, policies and strategy.

As we step into the 21st century, the WMD threat will become more manifest and more threatening to U.S. security interests. As a consequence, one should expect a growing national consensus on the need to rebuild a galvanizing vision of the United States' role in the world, one that leads international efforts to maintain stability in regions affected by the proliferation of weapons of mass destruction.

As one of the world's first proliferators, Saddam Hussein provided us a useful template to understand better the revolutionary changes underway in military relationships worldwide and to serve as a basis for shaping policies, strategy, and forces that will respond effectively to new dangers. For that, we should thank him.

The Proliferation of Knowledge and Technology

A useful perspective of warfare in the 21st century is described aptly by Alvin and Heidi Toffler in *War and Anti-War*.

The revolutionary new linkages between knowledge, wealth and war, according to the Tofflers, are shaping future conflict. As summarized in Table 1–1, the Tofflers find that the international power structure already has been transformed into three separate and overlapping tiers.[1]

Table 1–1
A Three-Tiered International System

Low Technical Complexity	Medium Technical Complexity	High Technical Complexity
First Wave	**Second Wave**	**Third Wave**
Clockspeed: Slow	Clockspeed: Fast	Clockspeed: Electronic
Symbol: Hoe	Symbol: Assembly line	Symbol: Computer
Agrarian civilization	Industrial civilization	Post-industrial civilization
Product of the Agricultural Revolution	Product of the Industrial Revolution	Product of the Informational Revolution
Primitive weapons: spear/sword	Standardized weapons from mass production	Precision-strike conventional weapons, integrated air/missile defenses, global command and control
Modern Example: Rwanda Implosion	Modern Example: Iraq in Gulf War	Modern Example: U.S. in Gulf War
Hand-to-hand combat	Range restricted, mass for firepower	Range, speed, lethality allow dispersed firepower

Evolving Threat

> Nuclear, biological, chemical weapons; ballistic and cruise missiles; satellites and satellite navigation receivers; technical support

---

[1]Alvin and Heidi Toffler, *War and Anti-War* (New York: Little, Brown, 1993), pp. 18–22.

Alvin and Heidi Toffler's analysis makes it clear that in 1996 the problem of future strategy is intellectual. Understanding the revolutionary changes that are being created by the potential for global proliferation in the Second Wave is the most daunting challenge. Since the flow of information cannot be controlled in the modern world, the dispersal of technologies and knowledge of nuclear, biological, and chemical weapons, and cruise and ballistic missiles, promises to be more widespread as time passes. Many of the resulting technical problems raised by WMD proliferation cannot be resolved except in a global framework. A part of the solution to proliferation is political and economic but the greatest challenge facing American planners is the development of workable operational military strategies that will respond effectively to counterproliferation tasking early in the 21st century.

The advent of Second Wave powers obtaining WMD, cruise and ballistic missiles, precision navigation, and space-based platforms for situational awareness, targeting, command-control-communications, and other support for terrestrial forces is transforming military relationships between the powerful and not-so-weak states worldwide. The truth of the matter is that current U.S. policies, doctrine, strategy, and operational concepts are becoming increasingly outdated. How can the military establishment be expected to reach its politically-prescribed objectives if it is precluded, by the enemy's threat or use of force, from using in-theater ports, airfields, and assembly areas and if aircraft carrier operations are placed in jeopardy by land-based missile threats?

The technology race in creating wealth is already well underway. A banner headline on the cover of *Business Week* announces that "technology is transforming the American economy into the most productive in the world." In the same issue, Compaq's chief executive, Eckhard Pfeiffer observes that "for us to keep growing, we have to put in place state-of-the-art technology, make constant progress and make constant improvements. If you stand still, you will fall behind."[2]

To be certain, America's economic competitors will not stand still either. The need for constant technological progress is well-recognized worldwide. If Alvin and Heidi Toffler are correct when they say that "the way we make war reflects the way we make wealth,"[3] we can expect the technical military complexity of potential adversaries to be increasing rapidly and in new directions. The proliferation of weapons of mass destruction, missiles, advanced conventional weapons, and use of spaceborne sensors and navigation devices will present a totally new warfare environment.

More than twenty states are reported to be building the scientific, technological, and production infrastructures and gathering the expertise necessary for the development of weapons of mass destruction. In the nuclear realm, there are more than a dozen states that either possess or can produce nuclear materials; about two dozen countries are considered capable of producing chemical weapons and about half that many are engaged in or

---

[2]As quoted in "Riding High: Corporate America Now Has An Edge Over Its Global Rivals," *Business Week* (October 9, 1995), p. 134.

[3]Toffler and Toffler, *War and Anti-War*, p. 3.

capable of producing biological agents. In the areas of missile technology and production, more than a dozen countries are currently capable and ten more are on the verge of such capability. The confirmed, probable, and suspected proliferating countries are shown in Table 1–2.

Yesterday's customer is today's supplier. Ten years ago, the Western countries were the suppliers. Increasingly the Third World can be considered self-sufficient in missile systems, and some states can produce precursor chemicals. Russia and the Ukraine can export materials, technology and dual-use items. They also provide the know-how: North Korea and China have benefitted from the Russian-Ukrainian brain drain.

Yet, the Western nations remain the preferred sources of WMD materials. The openness of Western culture and the push of global trade provide encouragement. The relative ease of circumvention of export controls, especially on dual-use commodities and technologies associated with weapons of mass destruction, provide additional incentives.[4]

On the demand side, it appears that the number of states capable of possessing WMD is increasing: five declared and

---

[4]The international community has several non-proliferation treaties and control regimes in place: Nuclear Non-Proliferation Treaty (renewed in 1995), Biological Weapons Convention (1975), Chemical Weapons Convention (1993; not yet in force), Missile Technology and Control Regime (1987), and the Australia Group (1984) which identifies for international control 54 precursor chemicals and biological agents. In addition, the Cold War-oriented Coordinating Committee on Multi-Lateral Export Controls (1949–1994) was replaced by an "Export Forum" in 1995 for the purpose of coordinating the international movement of dual-use technologies.

three more possibly nuclear capable. It also appears that most potential nuclear states will rely on aircraft as the preferred means of delivery, but they also will pursue cruise and ballistic missiles and can apparently afford to do so. It is clear, and probably to be expected, that more states are seeking chemical and biological weapons than nuclear weapons. They are easier both to produce and to conceal. Chemical and biological weapons are becoming the "poor man's atomic bomb."

Unsettling signals persist. Some of the more problematic proliferators continue to remain outside the established control regimes. Not all nations adhere to or honor all the provisions of international conventions and treaties. Of the nations signing the Biological Weapons Convention, for instance, many still may be conducting biological weapons research and many more nations continue to produce the chemicals considered to be precursors of chemical weapons.

Despite the international and technical hurdles for WMD programs, the proliferation of essential components continues. One reason for the ease of proliferation is that many of the raw materials needed to engage in WMD programs also have legitimate uses in industry and economic development programs—"dual-use" commodities and technologies. Detecting nuclear, biological, and chemical weapons programs, as well as missile and military satellite programs, becomes very difficult because of the dual-use nature of the constituent elements. Many of the relevant technologies are actually "multi-use" in the sense that they can be used across a range of basic civil and military

Table 1–2
Suspected Weapons of Mass Destruction Programs

	Nuclear	Biological	Chemical	Ballistic Missile
Afghanistan		•	•	•
Algeria	?			
Argentina				•
Brazil		•		•
Bulgaria				
Chile			•	
Cuba		•	•	•
Czech Republic				•
Egypt		•	•	•
Ethiopia			•	
Hungary				•
India	•	•	•	•
Indonesia			•	
Iran	•	•	•	•
Iraq	•	•	•	•
Israel	•	•	•	•
Laos		•	•	
Libya		•	•	•
Myanmar (Burma)			•	•
North Korea	•	•	•	•

Pakistan		●				●
Poland			●			●
Romania						●
Saudi Arabia						●
Slovak Republic				●		●
South Africa						
South Korea					●	●
Syria		●			●	●
Taiwan		●			●	●
Thailand					●	
United Arab Emirates						●
Vietnam		●			●	●
Yemen						●

Source: U.S., Congress, Office of Technology Assessment, *Proliferation of Weapons of Mass Destruction: Assessing the Risks* (Washington, D.C.: Government Printing Office, August 1993), pp. 15, 80, 82; John M. Collins, Zachary S. Davis, and Steven R. Bowman, *Nuclear, Biological, and Chemical Weapon Proliferation: Potential Military Countermeasures*, Congressional Research Service (Washington, D.C.: Library of Congress, June 28, 1994), p. 4; U.S., Senate, Committee on Governmental Affairs, *Proliferation in the 1990's*, 103d Cong., 1st Sess. (Washington, D.C.: Government Printing Office, March 24, 1993), pp. 64–66; and William J. Perry, Secretary of Defense, *Annual Report to the President and the Congress* (Washington, D.C.: Government Printing Office, March 1996), p. 220.

uses such as computing, metal-forming, and diagnostic testing.[5] As will be shown in the Iraqi case study below, a determined proliferator easily can disguise procurement networks by using fronts, intermediaries in other countries, and other deception measures to circumvent many export controls.

The globalization of commerce is spreading advanced technologies worldwide. Among the countries trying to integrate the U.S. Global Positioning System (GPS) satellite receivers into missiles and unmanned aerial vehicles, for example, are Russia, Germany, France, Israel, China, Iran, Pakistan, and Myanmar (Burma). Although designed for commercial exploitation, GPS receivers, when integrated properly with cruise and ballistic missiles, can offer proliferators greater accuracy for delivery of WMD warheads. This easy access to improved navigation and guidance is affordable and passive in obtaining position information accurate to within 100 meters. At the same time, the U.S. has slashed export controls on telecommunications technologies and computers, and the once-secure U.S. dominance in encryption products and hardware is eroding.

The export of both satellites that can take high resolution pictures and the pictures themselves is advocated by many in the U.S. aerospace industry. While conventional wisdom argues that civilian satellites and space programs represent peaceful technology, the truth of the matter is that many space technologies are inherently dual-use in nature. There is an especially high crossover between civil and military uses in

---

[5]U.S., Congress, Office of Technology Assessment, *Proliferation of Weapons of Mass Destruction: Assessing the Risks* (Washington, D.C.: Government Printing Office, August 1993), p. 16.

imagery systems, including space program subsystems and components. As civilian uses move toward higher resolution, faster data delivery, and widespread use of communications satellites, the military crossover effects will be even greater, especially if a country possesses ballistic missiles and WMD warheads in the ready.

Despite the relative ease of obtaining essential components and expertise, would-be proliferators, as illustrated in Table 1–3, face daunting technical hurdles for their nuclear, biological, and chemical programs. A considerable amount of time, money, and effort is necessary to build the infrastructure necessary to produce WMD and their delivery systems. Several of the steps below are expensive and time-consuming and may require outside expertise to bring them to fruition. In other areas, relatively easy steps are possible by applying the "secrets" of technology in innovative ways.

Iraq's Weapons of Mass Destruction

Destruction of Iraq's weapons of mass destruction was one of the key objectives of the 1991 Persian Gulf War. The Coalition's war planners believed that striking Iraq's weapons of mass destruction would reduce much of the threat to other regional states. General Norman Schwarzkopf, Commander-in-Chief of the U.S. Central Command (CINCCENT), issued a mission statement containing five items, including the following: "As early as possible, destroy Iraq's ballistic missile, NBC [nuclear, biological and chemical] capability." Six principal military objectives were promulgated to satisfy the mission in U.S.

Table 1-3
Technical Hurdles for Nuclear, Biological, and Chemical Weapon Programs

	Nuclear	Biological	Chemical
Nuclear materials or lethal agents production:			
Feed materials	Uranium ore, oxide widely available; plutonium and partly enriched uranium dispersed through nuclear power programs, mostly under international safeguards.	Potential biological warfare agents are readily available locally or internationally from natural sources or commercial suppliers.	Many basic chemicals available for commercial purposes; only some nerve gas precursors available for purchase, but ability to manufacture them is spreading.
Scientific and technical personnel	Requires wide variety of expertise and skillful systems integration.	Sophisticated research and development unnecessary to produce commonly known agents. Industrial microbiological personnel widely available.	Organic chemists and chemical engineers widely available.
Design and engineering knowledge	Varies with process, but specific designs for producing either of the two bomb-grade nuclear materials can be difficult to develop: • Separation of uranium isotopes to produce highly enriched uranium;	Widely published; basic techniques to produce known agents not difficult.	Widely published. Some processes tricky (Iraq had difficulty with tabun cyanation; succeeded at sarin alkylation; however, sarin quality was poor).

Design and engineering knowledge	• Reactor production and chemical processing to produce plutonium.		
Equipment	Varies with different processes, but difficulties can include fabrication, power consumption, large size, and operational complexity: • Electromagnetic separation equipment can be constructed from available, multiple-use parts; • Equipment for other processes is more specialized and difficult to buy or build.	Widely available for commercial uses. Special containment and waste treatment equipment may be more difficult to assemble, but are not essential to production.	Most has legitimate industrial applications. Alkylation process is somewhat difficult and is unusual in civilian applications. Special containment and waste treatment equipment may be more difficult to assemble, but are not essential to production.
Plant construction and operation	Costly and challenging. Research reactors or electric power reactors might be converted to plutonium production.	With advent of biotechnology, small-scale facilities now capable of large-scale production.	Dedicated plant not difficult. Conversion of existing commercial chemical plants feasible but not trivial.
Overall cost	Cheapest overt production route for one bomb per year, with no international controls, is about $200 million; larger scale clandestine program could cost	Enough for large arsenal may cost less than $10 million.	Arsenal for substantial military capability (hundreds of tons of agent) likely to cost tens of millions of dollars.

Continued on next page

Table 1–3 *Continued*

	Nuclear	Biological	Chemical
Overall cost	10 to 50 times more, and even then not be assured of success or of remaining hidden. Black-market purchase of ready-to-use fissile materials or of complete weapons could be many times cheaper.		
Weaponization: Design and engineering	Heavier, less efficient, lower yield designs easier, but all pose significant technical challenges.	Principal challenge is maintaining the agent's potency through weapon storage, delivery, and dissemination. Broad-area dissemination not difficult; design of weapons that effectively aerosolize agents for precision delivery challenging (but developed by U.S. by '60s).	Advanced weapons somewhat difficult, but workable munition designs (e.g., bursting smoke device) widely published.
Production equipment	Much (e.g., machine tools) dual-use and widely available. Some overlap with conventional munitions production equipment.	Must be tightly contained to prevent spread of infection, but the necessary equipment is not hard to build.	Relatively simple, closely related to standard munitions production equipment.

Source: U.S. Congress, Office of Technology Assessment, *Proliferation of Weapons of Mass Destruction: Assessing the Risks* (Washington, D.C.: Government Printing Office, August 1993), pp. 10–11.

Central Command (CENTCOM) Operations Order 91–001, dated January 17, 1991, including: "Destroy known nuclear, biological, and chemical (NBC) production, storage, and delivery capabilities."[6] General Schwarzkopf told his Army corps and division commanders that he considered Iraq's nuclear and chemical capability as one of the country's three centers of gravity, the others being the Republican Guard and Saddam Hussein himself.[7]

Immediately after the Gulf War, many military analysts were optimistic that this objective had been accomplished, since damage to the known facilities involved in these programs was substantial. As inspectors from the International Atomic Energy Agency (IAEA) and the United Nations Special Commission discovered, however, the Iraqi nuclear, biological, chemical, and missile programs were much more extensive than were believed originally, and they were not set back nearly as much as initially indicated by the Coalition's battle damage assessments.

The Persian Gulf War. The full extent of Iraqi WMD programs was not well known when Iraq invaded Kuwait in August 1990. The Defense Intelligence Agency's Automated Intelligence File (AIF) on June 1, 1990, listed only eleven nuclear, biological, and chemical targets: eight related to chemical and biological production and storage, and three dealing

---

[6]U.S., Department of Defense, *Conduct of the Persian Gulf War* (Washington, D.C.: Government Printing Office, April 1992), pp. 72–74.

[7]Michael R. Gordon and General Bernard E. Trainor, *The Generals' War: The Inside Story of the Conflict in the Gulf* (Boston: Little, Brown & Company, 1995), p. 157.

with basic and applied nuclear research and development. This list was to grow to a high of thirty-four targets by January 26, 1991; all of the additional targets were in the categories of chemical and biological production and storage categories. Scud-related targets numbered twenty-four on June 1, 1990, and grew to 123 by January 26, 1991, with eighty-six of the ninety-nine added installations being either fixed missile sites or fixed positions for mobile missile launchers. Table 1–4 indicates the various numbers of nuclear, biological, and chemical and Scud-related facilities included on the principal installation and target lists from June 1, 1990, through July 1, 1991. It is important to note that these lists were prepared in relative isolation from each other. It also must be remembered that the AIF is an installation file and not necessarily a target list, thus some of the facilities indicated may not be appropriate military targets.[8]

The U.S. and Coalition bombing campaign was unable to achieve the objective of degrading the threat from Iraqi nuclear, biological, and chemical programs and their delivery systems. The sheer number of installations involved in these largely clandestine programs was much greater than was known previously. On January 16, 1991, the CENTCOM Special Planning Group's (more commonly known as the Black Hole) list of nuclear-related targets included only the uranium mine at Al Qaim and the large complex at Al Tuwaitha. The Al Tuwaitha complex, also known as the Baghdad Nuclear Research Center (Tuwaitha is only 10 miles south of Baghdad), was well known to U.S. planners since the Israelis bombed the French made

---

[8]U.S., Department of the Air Force, *Gulf War Air Power Survey: A Statistical Compendium and Chronology*, Vol. V, Part 1 (Washington, D.C.: Government Printing Office, 1993), pp. 215–22.

Table 1–4
Variation in Number of WMD and WMD-Related
Installations

Source	Date	NBC	Scud
DIA AIF	6-1-90	11	24
U.S. Air Force Central Command (CENTAF) Exercise Internal Look 90	6-15-90	10	10
U.S. Central Command (CENTCOM)	6-27-90	9	1
CENTAF	8-8-90	7	1
Air Staff Instant Thunder	8-15-90	6	2
CENTAF	8-16-90	8	3
CENTAF	9-2-90	10	12
Master Target List	11-2-90	11	13
Master Target List	12-8-90	13	13
Master Target List	1-15-91	20	45
DIA AIF	1-26-91	34	123
Master Target List	1-28-91	14	31
Master Target List	2-12-91	14	27
Master Target List	3-1-91	26	48
DIA AIF	7-1-91	17	108

Source: U.S., Department of the Air Force, *Gulf War Air Power Survey: A Statistical Compendium and Chronology*, Vol. V, Part 1, Table 61 (Washington, D.C.: Government Printing Office, 1993), p. 222.

Osirak reactor there in 1981.[9]   Ironically, post-Gulf War disclosures by Dr. Jaffar Dhia Jaffar, the head of Iraq's uranium enrichment and weaponization program, indicate that the Israeli attack was a turning point for the Baghdad government since it triggered a switch from plutonium-oriented reactor technology to

---

[9]*Ibid.*, pp. 217–22.

a more diversified and easily concealed uranium enrichment program.[10]

In attempting to destroy Iraq's NBC capability, U.S. and Coalition forces flew 970 attacks against this target category. More than forty percent of these strikes were made with precision weapons, and about eighty percent of these were conducted by F-117s. The bulk of the attacks was focused on chemical warfare capabilities.[11]

Post-War Discoveries by U.N. Inspectors. Following the Gulf War, the United Nations adopted U.N. Security Council Resolution 687, which, among other matters, was directed at eliminating Iraq's WMD and the means to produce and use them. Inspections of suspected nuclear facilities were conducted by the IAEA, while the U.N. Special Commission on Iraq (UNSCOM) was tasked with inspecting biological, chemical, and missile facilities. While both groups carried out their assigned responsibilities aggressively, the IAEA has been much more open in revealing its findings in unclassified literature, submitting a total of twenty-eight reports to the U.N. through 1995 with numerous details of the Iraqi nuclear program.[12] The Special Commission considers the reports of the on-site inspection teams

---

[10]Barbara Ebert, "Iraq: Its Nuclear Past As a Way of Assessing Its Nuclear Future," unpublished report (Vienna, Va.: Science Applications International Corporation, April 1994), p. 5.

[11]U.S., Department of the Air Force, *Gulf War Air Power Survey: Summary Report* (Washington, D.C.: Government Printing Office, 1993), p. 80.

[12]United Nations, Security Council, *Report of the Twenty-Seventh IAEA On-Site Inspection in Iraq Under Security Council Resolution 687 (1991)*, S/1994/1443 (December 22, 1994), p. 3.

to be sensitive since they are an integral part of the still ongoing investigation. Hence, its reports indicate that it has inspected "X" number of installations, but few names or locations are revealed, and most details concerning these installations have been kept behind closed doors.[13] As a result, alternative resources were used to compile the installation information for chemical and biological facilities. These included reports compiled by the Gulf War Air Power Survey, reports for Congress by the Congressional Research Service, and recent books and articles on Iraq's military capabilities and about the Gulf War. Emerging from this research approach is a mosaic of Iraq's WMD and missile programs, one that points the way ahead for other would-be proliferators.

The success and enormous complexity of the Iraqi research and development and production efforts can serve as an analytical pivot to address several daunting proliferation challenges that will face U.S. operational planners in the years ahead. The possession and potential use of nuclear, biological, and chemical weapons by Third World countries—or "Second Wave" states—requires a significant shift in American thinking about how best to approach future contingencies.

---

[13]Telephone interview with Mr. Charles Duelfer, Deputy Chairman, U.N. Special Commission on Iraq (May 30, 1995).

Chapter 2
Building a Nuclear Arsenal

Writing for *The Bulletin of the Atomic Scientists* in March 1991, just after the Persian Gulf War, David Albright and Mark Hibbs summarized the conventional wisdom at the time about Iraq's nuclear weapons.

Confusing reports in the early days of the war suggested that Iraq had extensive nuclear weapons facilities and Pentagon secrecy about bombing targets added to the confusion. But Iraq's nuclear weapons research and related activities are almost certainly carried out at only a few sites, and the major ones are largely devoted to other activities.... The war may put a damper on any future program by damaging Iraq's industrial capacity, but the nuclear effort was at such an early stage that there was little to destroy.[1]

These views generally were shared by the military. In June 1990, a U.S. Central Command "Iraqi Target Study," for example, contained one known nuclear target. Military planners believed that the Iraqi nuclear weapons program had not gone beyond research. Intelligence reports indicated in November 1990, three months after Baghdad's takeover of Kuwait, that the Iraqis had launched a "crash program" to fabricate one or two "crude nuclear explosive devices" in six months to a year.

---

[1]David Albright and Mark Hibbs, "Iraq and the Bomb: Were They Even Close?," *Bulletin of the Atomic Scientists* (March 1991), pp. 24–25.

Nonetheless, U.S. target analysts insisted that any devices resulting from these efforts would have limited usefulness. Based on what was believed to be the state of the Iraqi program at the outset of Operation Desert Shield in August 1990, military planners were convinced that Iraq's hurried efforts would result in devices of dubious reliability and low yield that would lack appropriate delivery vehicles. When U.S. and Coalition air attacks were initiated on January 16, 1991, only two nuclear targets were included in the first day's air tasking order.[2]

The Persian Gulf War

During the Gulf War air attacks from January 16 through February 28, 1991, U.S. and Coalition fighter-bombers repeatedly struck the number one Iraqi nuclear facility on the target list—the Baghdad Nuclear Research Center at Al Tuwaitha. American aircraft returned often to pound Iraq's premier nuclear facility. By the end of the war, seven more targets suspected of having a nuclear role were included on the Coalition target lists. American intelligence only had begun to realize that the Iraqi nuclear program extended far beyond Al Tuwaitha. In the last days and hours of the war, as shown in Figure 2–1 and Table 2–1, Al Tuwaitha and the seven suspected nuclear facilities were hammered from the air.

---

[2]U.S., Department of the Air Force, *Gulf War Air Power Survey: Summary Report*, pp. 32, 43, 79 and *Gulf War Air Power Survey: Planning and Command and Control*, Vol. I, Part 1 (Washington, D.C.: Government Printing Office, 1993), pp. 159–60, 189, 193.

Figure 2–1
Eight Nuclear Targets Bombed During the Gulf War

Table 2–1
Eight Known and Suspected Nuclear Facilities During the Gulf War
(last 72 hours of the war)

Nuclear Facility	Location	Bomb Damage
Al Jesira	25 mi NW of Mosul	Twenty-five air strikes, mostly in the last two weeks of the war; multiple 2,000-pound and 500-pound bomb hits.
Al Atheer	35 mi SW of Baghdad	Coalition targeters learned about its nuclear role late in the war; only a few days to bomb before war's end.
Al Qaim	235 mi NW of Baghdad	Multiple air strikes resulting in major damage.
Ash Sharqat	125 mi NW of Baghdad	Several raids in mid-February against this "rocket facility;" intelligence did not suspect a nuclear connection.
Tarmiya	25 mi N of Baghdad	Regarded as a rocket facility; upgraded to a "possible" nuclear facility on February 23.
Al Tuwaitha	10 mi S of Baghdad	Bombed often throughout the war; Iraq's premier nuclear target.
Underground Facility	30 mi N of Mosul	One strike with four laser-guided bombs with penetrating warheads; all missed.
Not Identified	8 mi S of Baghdad	10 bombs on target on February 22.

Source: U.S., Department of the Air Force, *Gulf War Air Power Survey: Operations and Effects and Effectiveness*, Vol. II, Part 1, pp. 225–27 and *Gulf War Air Power Survey: A Statistical Compendium and Chronology*, Vol. V (Washington, D.C.: Government Printing Office, 1993), p. 218.

Gulf War planners were faced with the reality that Iraq's clandestine nuclear program was designed from the start to secret its bomb-making efforts from the IAEA and U.S. spy satellites.

Although it may be shocking to Western nuclear weapons experts, the Iraqis turned to "old" technology. Iraqi scientists essentially replicated the technologies at the Oak Ridge Plant in Tennessee which helped to produce the first U.S. atomic bombs. The Iraqis built huge antiquated machines called "calutrons" for enriching uranium to weapons-grade levels. An elaborate complex of nuclear research and development facilities was established in various parts of the country. This effort was supported by 2,000 foreign-trained scientists, 18,000 engineers, and numerous Jordanian front companies to obtain needed materials from abroad. Evidence of these activities in Western intelligence was fragmentary. After the war, the Defense Intelligence Agency concluded that "prior to Desert Storm, little was known about Iraq's highly compartmented nuclear weapons program."[3]

To make matters worse for the Coalition's target planners, the Iraqis spirited away much of the program's critical machinery to open fields or areas near known nuclear installations like Al Tuwaitha. In essence, the Iraqi fixed nuclear targets were transformed into relocatable ones. American bombs often hit empty buildings or those whose critical components had been removed. The U.S. Air Force concluded:

> Bombing alone, therefore, failed to achieve the objective of *eliminating* the existing Iraqi nuclear weapons program. The Iraq nuclear program's redundancy, advanced status on the eve of the war, and elusiveness, in conjunction with the extraordinary measures the Iraqis

---

[3]As quoted in Gordon and Trainor, *The Generals' War*, pp. 181–82, 457.

took immediately after Desert Storm to conceal its extent by destroying certain facilities, all argue that the air campaign no more than 'inconvenienced' Iraqi plans to field atomic weapons.[4]

With Western intelligence agencies thrown totally off track by Iraq's deception measures, the Coalition's military planners operated on the basis of what were to be proven to be faulty assumptions. These wrong operating premises, when combined with the Iraqis' persistent boldness in doing the unthinkable, like moving critical machinery out of buildings and into the open field or adjacent structures, resulted in the Coalition's counter-nuclear operations being conducted in the blind.

IAEA inspection teams operating under U.N. Security Council Resolution 687 soon discovered a vast clandestine nuclear weapons program representing an investment of several billion dollars. David Kay, a member of the IAEA Action Team assembled to deal with Iraq led four inspection teams that uncovered Iraq's clandestine nuclear weapons program. He says:

They were probably 36-months away from having deliverable type of weapons that would have fit either their own Scud missiles or extended range Scuds, or in their high performance aircraft in small numbers. If the war had not intervened—as we sit here today in mid-1994, they would have at least had crude nuclear devices,

---

[4]U.S., Department of the Air Force, *Gulf War Air Power Survey: Operations and Effects and Effectiveness*, Vol. II, Part 2 (Washington, D.C.: Government Printing Office, 1993), p. 329.

and probably their first deliverable packages would have [been] coming on line.[5]

Iraq's Nuclear Weapons Program

The United Nations Security Council Resolution 687 was adopted on April 3, 1991, setting the terms for a formal ceasefire in the Gulf War. Under the terms of the Resolution, Iraq was to declare all locations, amounts, and types of nuclear weapons materials, subsystems, components, or research, development, support, or nuclear production facilities. To ensure compliance, Iraq was compelled to place all nuclear weapons materials under the exclusive control of IAEA and accept on-site inspection and destruction, removal, or rendering harmless all such items.[6]

Petrochemical Project 3 (PC 3). When the Iraqis decided to embark on a clandestine nuclear weapons program while Iraq was still a member of the Non-Proliferation Treaty, they created a front organization with the code name "PC 3." The broad-based effort to design and develop implosion-type weapons involved three ongoing uranium enrichment programs, a massive foreign procurement program shrouded behind an array of sophisticated deception practices, foreign technology assistance, and an investment of some ten to twelve billion dollars. The highly

---

[5]As quoted in a 15-minute report entitled "Saddam's Secret Bomb," London ITV Television Network (1830 GMT, May 19, 1994) in U.S. Foreign Broadcast Information Service, JPRS Report, *Proliferation Issues*, JPRS-TND-94–013 (June 24, 1994), pp. 23–24.

[6]United Nations, *Security Council, Resolution 687* (1991), U.N. Doc. S/RES/687 (1991) (April 3, 1991).

compartmentalized program was operated through an interlocking relationship between the Iraqi Atomic Energy Commission (IAEC) and the Ministry of Military Industry and Industrialization.

Many Western scientists and nuclear experts now believe Iraq was twelve to eighteen months away from producing a nuclear bomb. "When it comes to nuclear weapons," one American recalls about pre-Gulf War U.S. intelligence reports, "it's easy to dismiss a country like Iraq." An Israeli expert called the inability of Western governments to detect PC 3 a "colossal intelligence failure." Hans Blix, director of IAEA which conducted inspections of Iraq's declared nuclear facilities before the Gulf War observes that "there was suspicion certainly [but] to see the enormity of it is a shock."[7]

PC 3 was a complex, comprehensive nuclear weapons development program characterized by parallel approaches to fissile material production and by theoretical and experimental design work. The Iraqis were pursuing three industrial-scale methods of uranium enrichment: electromagnetic isotope separation (EMIS), chemical processes, and gas centrifuge. A fourth approach, the gaseous diffusion method, was discarded in the 1980s after the Iraqis concluded it was too risky. At the R&D level, at least as far as the IAEA inspectors know, the Iraqis also

---

[7]As quoted in R. Jeffrey Smith and Glenn Frankel, "Saddam's Nuclear-Weapons Dream: A Lingering Nightmare," *Washington Post* (October 13, 1991), pp. A1, A44, A45.

tried Atomic Vapor Laser-Isotope Separation (AVLIS) and Molecular Laser Isotope Separation (M-LIS).[8]

The nuclear program was supported by broad-based international procurement efforts. For instance, IAEA inspectors found numerous catalogues from foreign suppliers. The Iraqis devised a multitude of plausible cover explanations for purchases from overseas. While the Ministry of Industry and Military Industrialization maintained general control over the weapons program, PC 3 was assigned specific control of nuclear weapons.[9]

The Al Tuwaitha Nuclear Research Center was the nerve center of the Iraqi clandestine weapons complex. It was made up of more than 100 buildings and included research reactors, hot

---

[8]The most difficult process of producing nuclear weapons is production of highly enriched uranium-235 (HEU). This material can sustain nuclear chain reactions that release tremendous amounts of energy in a short time. To manufacture HEU for weapons, the uranium-235 isotope must be separated from the more common uranium-238. The various enrichment methods differ in terms of the number of stages required to enrich uranium to weapons grade levels (some 90 percent or higher). Each stage of enrichment takes an input of uranium, or "feed," and produces a greater concentration of uranium-235 (the "product") and depleted uranium-235 (the "tails"). In other methods, however, each stage accounts for only marginal enrichment; many successive stages are required to obtain HEU through "cascades" where each successive stage results in the next higher level of enrichment and the tails are left behind (gaseous diffusion, gas centrifuge, and chemical processes). See U.S., Office of Technology Assessment, *Technologies Underlying Weapons of Mass Destruction* (Washington, D.C.: Government Printing Office, 1993), pp. 120–21, 140–43.

[9]United Nations, Security Council, *First Report on the Sixth IAEA On-Site Inspection in Iraq Under Security Council Resolution 687 (1991)* (September 22–30, 1991), U.N. Doc. S/23122 (October 8, 1991), pp. 3–7.

cells, radiochemistry and radioisotope production laboratories, workshops, radioactive waste treatment facility, a material testing laboratory (ceramics and metals), nuclear physics laboratories, a complex of chemistry and chemical engineering research and development, and other nuclear weapons-related activities. Many of the buildings were destroyed during the Gulf War and two functioning research reactors were destroyed or heavily damaged.[10]

While only eight targets were identified during the Persian Gulf War, the target deemed by U.S. planners as the most important nuclear facility in Iraq was the Baghdad Nuclear Research Center in Al Tuwaitha. A sketch of the installation, prepared for U.N. and IAEA inspectors, is shown in Figure 2–2. In mid-February 1991, General Schwarzkopf directed Lieutenant General Charles Horner, Commander of the Air Force Component, U.S. Central Command (CENTAF), to develop a seventy-two hour bombing campaign against any remaining nuclear, biological, and chemical targets "in the event a ceasefire is declared and only seventy-two hours remain prior to plan implementation." The Al Tuwaitha complex headed General Horner's list, the same position it had occupied since the crisis began in August 1990.[11] The bombing devastated the site and much of the installation's equipment was destroyed. Table 2–2

---

[10]United Nations, Security Council, *Consolidated Report On the First Two IAEA Inspections Under Security Council Resolution 687 (1991) of Iraqi Nuclear Capabilities* (July 11, 1991), U.N. Doc. S/22788 (July 15, 1991).

[11]Gordon and Trainor, *The Generals' War*, pp. 224–25.

Figure 2–2
Al Tuwaitha

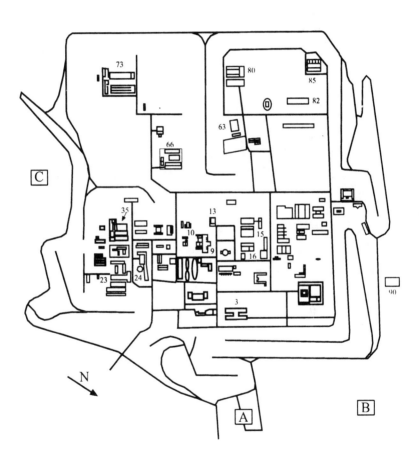

Table 2–2
Al Tuwaitha Buildings Involved in
Weaponization and Enrichment

Building No.	Description of Building	Activity
3	Administration Building	Personal computers for hydrodynamic models.
9	Chemical & Radiochemical Analysis Laboratory	Separation of plutonium from exempted pins. Separation of plutonium from irradiated pins.
10	Chemical Analysis Laboratory	Production of uranium metal. Melting and casting of metal uranium.
10 annex	Nuclear Physics Department	
13	Research Reactor IRT-5000	Irradiation of cassettes.
15	Isotope Production Laboratory	Production of UF_4 and UF_6.
16	Workshop	Initiator workshop.
23	Laboratory Workshop Building	Gaseous diffusion enrichment. Ceramic capacitor fabrication.
24	Tamuz-2 Research Reactor	Storing of irradiated cassettes. Disassembling of cassettes. Neutron measurements.
35	Radioactive Waste Treatment Station	Handling of wastes from the irradiation program.
63	Cold Water Testing Laboratories	Gas centrifuge enrichment.
66	Training Building	Initiator system examination.
73	Experimental Fuel Fabrication Laboratories	Manufacturing of fuel.
80	Nuclear Physics Laboratories	EMIS.
82	Electronic Research Laboratories	Electronic systems.
85	Chemical Research Laboratories	Production of yellowcake, UO_2 and UCl_4.
90	Polymer Chemistry Research Laboratory	Enrichment by solvent extraction and ion exchange.

indicates the principal buildings at Al Tuwaitha that were involved in the enrichment and weaponization programs.[12]

The Iraqis apparently quickly concluded that much of the critical material and specialized equipment at this and other sites was vulnerable to U.S. and Coalition bombing. The Iraqis moved many of these items, such as nuclear fuel or critical machinery, to fields or other areas in the vicinity of known installations and buried them.

At Al Tuwaitha, for example, the Iraqis moved some irradiated fuel assemblies that had been stored there to a site about three miles north of the complex. This site, located in the Garf al Naddaf district, was named "Location B" by IAEA inspectors. It consisted of approximately an acre of soft clay farmland, with no built structures, no support fixtures such as water or electricity, and no roads leading to it. The Iraqis simply buried sixteen concrete storage tanks and filled them with cooling water. The irradiated fuel assemblies were placed on aluminum racks and put into carbon steel drums, with up to two drums stored in each tank. The tanks were buried below surface level, and were covered with a reinforced concrete plate, with a hole in the center for water refilling. This hole then was covered with a second, smaller concrete plate. A total of 132 fuel assemblies of different types were stored in this manner.[13]

---

[12]United Nations, Security Council, *Report on the Seventh IAEA On-Site Inspection in Iraq Under Security Council Resolution 687 (1991)* (October 11–22, 1991), U.N. Doc. S/23215 (November 14, 1991), p. 12.

[13]Fernando Lopez Lizana, Robert Ouvard, and Ferenc Takats, "Nuclear Inspections in Iraq: Removing Final Stocks of Irradiated Fuel," *IAEA Bulletin*, 3/1994, pp. 25–26.

While such methods of storing nuclear materials would be unheard of by Western standards, their use here illustrates the lengths the Iraqis went to salvage the materials and equipment they had spent a fortune buying or developing. Parts of huge calutron equipment, devices to accelerate uranium ions to high energy within a vacuum system in order to collect the uranium-235 and uranium-238 ions after they have been moved between the poles of a very large electromagnet, were discovered buried in the desert, and in some cases were loaded onto trucks and hauled to various locations in an attempt to elude IAEA inspectors.[14]

IAEA On-Site Inspections. The IAEA's first inspection was conducted in May 1991. It covered the nuclear research facility at Al Tuwaitha and another site in the Baghdad area. The main purpose was to verify the Iraqi's April 27 declaration concerning these sites. The inspectors found parts of the Al Tuwaitha complex to be heavily damaged from the bombing. The Iraqis, however, already had cleared much of the debris. In addition, documentation and records of activities at the facility had disappeared, and much of the equipment that once had existed at Al Tuwaitha had been removed to undisclosed locations.[15]

Twenty-eight IAEA nuclear inspection teams visited Iraq through the end of 1995 to inspect facilities, interview key personnel, inventory nuclear materials, identify prohibited items,

---

[14]David Albright and Mark Hibbs, "Iraq's Nuclear Hide-And-Seek," *Bulletin of the Atomic Scientists* (September 1991), pp. 14–22.

[15]United Nations, Security Council, *Consolidated Report on the First Two IAEA Inspections Under Security Council Resolution 687 (1991) of Iraqi Nuclear Capabilities* (July 11, 1991), p. 5.

and carry out destruction and removal operations. The inspectors found a large and complex clandestine nuclear weapons program carried out at ten dedicated sites (Table 2–3). Each site had several major buildings, representing a very large investment—millions of dollars worth of specialized equipment.[16]

Iraq's nuclear program is shown in Figure 2–3. In addition to the ten core sites, the IAEA included nine sites encompassing three main areas: uranium mining, production, and processing; uranium enrichment; and weaponization. Al Tuwaitha was involved in all three areas and Al Jesira was involved in the uranium production and enrichment processes. The massive industrial enterprise supporting the nuclear program involved many more locations. The number of facilities was higher than what might be expected because of the extraordinary complexity and compartmentation of the clandestine program.

Uranium Mining, Production, and Processing Sites. The six main locations supporting the preparation of uranium for the enrichment processes are depicted in Figure 2–3. As shown in Table 2–4, the Geological Survey Institute in Baghdad and the Abu Sukhayr Mine should be added to the list. This institute developed the process and pilot plant operations to recover uranium from carbonate ore from the Abu Sukhayr Mine.

Uranium ore commonly is mined along with other mineral-bearing ores; only one part in 500 is uranium. Milling facilities

---

[16]International Atomic Energy Agency, *IAEA Inspections and Iraq's Nuclear Capabilities* (Vienna, Austria: April 1992), pp. 5–13.

Table 2–3
Iraq's Ten Core Nuclear Facilities

Nuclear Site	Description
Al Tuwaitha Nuclear Research Center	Main site for Iraqi nuclear program. Activities included: several research reactors, plutonium separation and waste processing, uranium metallurgy, neutron initiator development, and work on a number of methods of uranium metalullurgy, neutron initiator development, and work on a number of methods of uranium enrichment.
Tarmiya	Main site for EMIS program for the enrichment of uranium. Site included both 1200 mm and 600 mm separators. Much of the equipment was disassembled by the Iraqis and the components hidden from IAEA inspection teams.
Al Atheer Center	This site was designed and constructed as the major facility for nuclear weapons development and testing. Activities at the site included: uranium casting and metallurgy, core assembly, explosive lens assembly, and detonics testing. A high explosives test bunker near the site was used for hydrodynamic experiments.
Al Furat	This site was intended for the design, assembly, and testing of gas centrifuges for uranium enrichment. A 100 centrifuge cascade was planned.
Al Jesira factory	Uranium feed stock production facility. Products included uranium dioxide, uranium tetrachloride, and uranium hexaflouride.
Akashat Mine	Uranium ore production site; associated with Al Qaim site.
Al Qaim	Production of yellowcake (refined uranium ore); ore supplied by foreign and domestic sources.
Rashdiya	Centrifuge development center; engaged in centrifuge design and testing.
Ash Sharqat	Site intended as a duplicate of the Al Tarmiya EMIS facility.
Petrochemical -3 Center	Complex of five office buildings in central Baghdad housing Iraqi nuclear weapons design effort.

Source: United Nations, Security Council, Special Commission on Iraq, information paper (October 16, 1995).

Figure 2–3
Nineteen Primary Sites Related to the Iraqi Nuclear Weapons Program

Uranium mining, production, processing sites

- Al Tuwaitha Nuclear Research Centre
- Al Jesira
- Mosul (processng)
- Al Qaim
- Tikrit (yellowcake storage)
- Akashat (phosphate & uranium processing)

Sites related to uranium enrichment

- Al Tuwaitha Nuclear Research Centre
- Al Tarmiya (EMIS)
- Ash Sharqat (planned)
- Al Jesira
- Al Radwan
- Al Amir
- Al Furat (centrifuge production)
- Daura (manufacturing)
- Badr (engineering complex)
- Salladine
- Nassar Works

Sites related to weaponization

- Al Tuwaitha Nuclear Research Centre
- Al Atheer
- Al Qa Qaa
- Hatteen (high explosives test site)
- Al Hadre (high explosives test site)

△ Al Jesira
● Mosul
△ Al Hadre
△ Ash Sharqat
△ Tikrit
△ Salladine
Al Tarmiya △
Al Ameen ★ BAGHDAD
Al Radwan △ △
Al Amir △ Daura △ Al Tuwaitha
△ Al Furat
Al Qa Qaa △ △ Badr
Hatteen △ △ Al Atheer
△ Al Qaim
△ Akashat

Source: International Atomic Energy Agency, *IAEA Inspections and Iraq's Nuclear Capabilities* (Vienna, Austria: April 1992).

Table 2–4
Uranium Mining, Production, and Processing Sites

Name	Also Known As	Location	Purpose
Abu Sukhayr	Abu Skhair, Abou-Sukhair	15 mi SW of Najar	Carbonate ore mine. Uranium in ore ranged from 80 to 800 ppm. Prospecting stopped in mid-1990 after mine shaft flooded. Discovered after the war.
Akashat	Ahasat		Iraq's main uranium mine. Mined uranium-bearing phosphorous rock.
Al Jesira feed materials plant	Al Jasira, Jazirah, Al Jezira	Outside Mosul	Converted yellowcake to nuclear grade UO_2 and UCl_4 for the EMIS program. Under consideration for future production of UF_6. EMIS production, centrifuge production. Destroyed in Gulf War.
Al Mosul		Outside Mosul	Processing.
Al Qaim	Al Qa'im	235 mi NW of Baghdad	Uranium concentration plant. Uranium was recovered from phosphate ore at a yellowcake production facility collocated with a super-phosphate fertilizer plant. Destroyed in Gulf War.
Al Tuwaitha Nuclear Research Center	Baghdad Nuclear Research Center	Outside Baghdad	Technical headquarters of the Iraqi nuclear program and site of many R&D functions. Approximately 100 structures in the compound. Mostly destroyed in Gulf War.
Geological Survey Institute		Baghdad	Process development and pilot plant operations to recover uranium from carbonate ore from Abu-Sukhayr Mine.
Tikrit			Storage of yellowcake.

extract the uranium concentrate known as "yellowcake." The next step in the process is to convert the yellowcake to a form of uranium suitable for enrichment.[17] The other major Iraqi nuclear installation that was known to U.S. planners at the start of Desert Storm was the Al Qaim uranium concentration plant approximately 235 miles northwest of Baghdad, near the Syrian border. This plant was destroyed during the Gulf War.

Another major facility involved in uranium processing was the Al Jesira feed materials plant, outside Mosul in northern Iraq. This site produced uranium tetrachloride (UCl_4), a feed material for the EMIS enrichment process.   This plant was also substantially destroyed by Gulf War air strikes.[18] The sketch of this facility used by IAEA is at Figure 2–4.

The chemical engineering laboratory at Al Tuwaitha (Building 85) provided the feed material for the initial start up of Tarmiya. A new feedstock plant at Al Jesira near Mosul would support full-scale production work at Tarmiya and, when it was completed, Ash Sharqat. Two separate plants were at Al Jesira, one for uranium oxide (UO_2) production and one for UCl_4. According to Iraq's authorities, two additional processes were

---

[17]U.S., Congress, Office of Technology Assessment, *Technologies Underlying Weapons of Mass Destruction*, pp. 137–38.

[18]Ebert, "Iraq: Its Nuclear Past As a Way of Assessing Its Nuclear Future," p. 5, and IAEA "Fact Sheet" (January 1994) enclosed in a letter to the author from Maurizio Zifferero, Leader of UNSC 687 Action Team (Vienna, Austria: International Atomic Energy Agency, January 18, 1994), p. 3.

Figure 2–4
Al Jesira

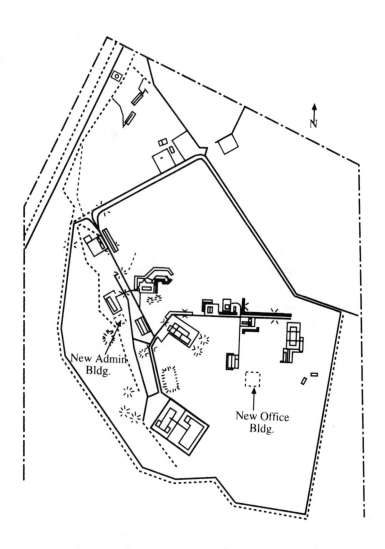

scheduled for the Al Jesira plants, including a uranium hexaflouride (UF_6) production plant for the centrifuge program.[19]

Sites Related to Uranium Enrichment. Eleven main sites for enrichment of uranium are indicated in Figure 2–3. Owing to the compartmentation of various elements of the Iraqis' uranium enrichment program, many more manufacturing facilities that were supporting the effort have been identified. Table 2–5 identifies twenty-four sites involved in the process.

The Iraqis were developing simultaneously three methods of uranium enrichment, including gas centrifuge, chemical processes from France, and electromagnetic isotope separation. The research and development activities for the Iraqi centrifuge enrichment program were conducted at Al Tuwaitha along with some manufacturing tests at Al Furat. Machinery of required parts took place at the Saddam Works, the Nassar Works, and the Rashdiah Engineering Center for centrifuge research and development. These were to be used at Al Furat, which was planned as the center for centrifuge assembly and testing.[20]

Iraqi scientists also were experimenting with French-developed techniques that would yield weapons-grade uranium. These techniques involved the liquid-liquid solvent extraction (Chemex) process. Previously, the Iraqis experimented with the

---

[19]United Nations, Security Council, *Report on the Fourth IAEA On-Site Inspection in Iraq Under Security Council Resolution 687 (1991)* (July 27-August 10, 1991), U.N. Doc. S/22986 (August 28, 1991), pp. 7–8.

[20]United Nations, Security Council, *Report of the Seventh IAEA On-Site Inspection in Iraq*, p. 23.

Table 2–5
Sites Related to Uranium Enrichment

Name	Also Known As	Location	Purpose
Al Ameen			EMIS component manufacturing.
Al Amil	Project 7307	4 mi W of Tarmiya	Liquid nitrogen plant—provided liquid nitrogen for EMIS diffusion pumps at Tarmiya.
Al Amir	Amir, Al Ameer Factory, Saddam State Establishment	Near Falluja	Calutron component manufacturing: magnet cores, return irons, ion sources and collector parts.
Al Furat	Al Kadisya State Establishment	18 mi SW of Baghdad, close to Badr Engineering Complex, An Walid	Centrifuge manufacturing facility where Iraq had planned to establish its first pilot cascade of 100 centrifuges. Centrifuge R&D, production and component manufacturing. Four main buildings.
Al Hamath		Al Tuwaitha (outside berm)	Workshop. U.N. concluded site was magnet test facility. Undamaged in Gulf War.
Al Nassariya	Ur State Establishment		Aluminum tubing for centrifuge molecular pumps.
Al Rabiyah Manufacturing Plant	Za'afaraniya Production Facility, Al Rabee, Al Nida, Al Nidda	at Za'afaraniya	Manufacturing plant with large mechanical workshops for the manufacture of large metal components for the Iraqi EMIS program. Damaged by cruise missile attack 17 Jan 1993; totally rebuilt by June 1993.
Al Radwan			EMIS component manufacturing.
Al Shaheed Brass Factory		Baghdad	Produces copper and brass ingots, rods, disks, tubing and wire. Does not produce finished products.

Al Shaheed Brass Factory			Capable of producing copper tubing of type used in EMIS coil windings.
Al Tuwaitha Nuclear Research Center	Baghdad Nuclear Research Center	10 mi S of Baghdad	Technical headquarters of the Iraqi nuclear program and site of many R&D functions. About 100 structures in the compound. Mostly destroyed in Gulf War.
Ash Shakyli	Al Shaykili; Al Shaykhili	Al Tuwaitha	Warehouse 13b storing centrifuge components.
Ash Sharqat	Al Sharqat	125 mi NW of Baghdad	EMIS production plant. Twin of Tarmiya that had not been completed. Heavily bombed in Gulf War. Three geographically separate locations 1–2 km apart.
Badr	Al Meelah Center		Centrifuge component manufacturing: vacuum chamber parts; engineering complex.
Daura	SEHEE (State Establishment for Heavy Engineering Equipment)		EMIS component manufacturing, Centrifuge component manufacturing.
Garf Al Naddaf	Location B	3 mi N of Al Tuwaitha Nuclear Research Center	Irradiated Fuel Storage. Nuclear fuel assemblies stored here to prevent destruction from bombing. No built structures, support facilities, water or electricity. Sixteen concrete storage tanks buried and filled with water. No damage in Gulf War.
Nahrawan	Al Nahrawan	12 mi E of downtown Baghdad	Conventional ordinance (munitions) plant. Location of computer numerically controlled (CNC) machine tools. Badly damaged in Gulf War.
Nassar Works	Nasser Works		Centrifuge component manufacturing.

Continued on next page

Table 2-5 *Continued*

Name	Also Known As	Location	Purpose
Nassr State Establishment	Nassr State Enterprise	Taji	Machine tools usable in nuclear program.
Saddam Works			Centrifuge component manufacturing.
Salladine	Salah Al Din State Establishment; SAAD-13	Salad Al-Din province, 15 mi N of Samarra	EMIS component manufacturing. Possible site of underground plutonium reactor. Badly damaged in Gulf War. U.N. concluded it unlikely an underground reactor was constructed at site.
Schuala	Al Schaula		Subsidiary establishment of Nassr State Establishment at Taji. Contains machine tools capable of involvement in the nuclear program.
Taji Center For Metallurgical Industries	Taji Metallurgy Laboratory	Taji	Studying industrial problems in materials science such as copper production, lead purification, refractory compounds.
Tarmiya	Al Tarmiya; Tamarya	25 mi N of Baghdad	ElectroMagnetic Isotope Separation (EMIS) plant for R&D & production to support industrial scale enrichment operations.
Um Al Maarik	Um Al Maarick; Auqba bin Nafi State Establishment; Um Al Marik	Badr	Contains machine tools and other relevant equipment that must be monitored. Contains four large diameter milling and boring machines located in building 119—used to make EMIS poles. Distinct group of buildings at Badr.

Japanese Asahi ion-exchange process but abandoned it after judging it too risky.[21]

The Tarmiya EMIS plant, located approximately twenty-five miles north of Baghdad, was designed by the Iraqis to support industrial-scale uranium enrichment operations. The installation housed two EMIS operations and was largely destroyed during the Gulf War. At the time it was bombed, however, U.S. intelligence regarded it as a rocket facility. The discovery late in the Gulf War that it could be a possible nuclear facility warranted additional strikes. Bad weather, however, caused several sorties to be less-than-effective or to be scrubbed altogether. As a result, Tarmiya ended the war as the top target on the Black Hole's priority list. The sketch of this large complex used by IAEA is shown at Figure 2–5. A twin of the Tarmiya plant was still under construction at Ash Sharqat, 125 miles northwest of Baghdad. This installation also was bombed during the Gulf War, and sustained heavy-to-moderate bomb damage.[22]

A principal facility for Iraq's gas centrifuge program, as far as we know at the end of 1995, was under construction at Al Furat, approximately twenty miles southwest of Baghdad. While the

---

[21]U.S. Congress, Office of Technology Assessment, *Technologies Underlying Weapons of Mass Destruction*, pp. 151, 178.

[22]Ebert, "Iraq: Its Nuclear Past As a Way of Assessing Its Nuclear Future," pp. 6–7; IAEA, "Fact Sheet," p. 3; U.S., Department of the Air Force, *Gulf War Air Power Survey: Operations and Effects and Effectiveness*, Vol. II, Part 1 (Washington, D.C.: Government Printing Office, 1993), p. 228; and United Nations, Security Council, *Report on the Twelfth IAEA On-Site Inspection in Iraq Under Security Council Resolution 687 (1991)* (May 26-June 4, 1992), U.N. Doc. S/24223 (July 2, 1992), p. 30.

**Figure 2–5
Tarmiya**

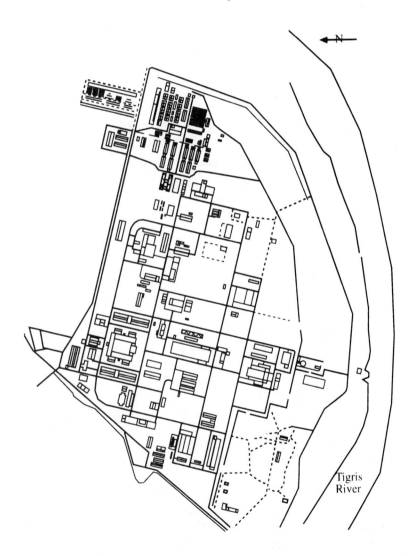

centrifuge program was second in priority to the EMIS program and was started much later, Iraqi authorities realized its potential and aggressively pursued obtaining the equipment necessary for this program, which relied heavily on imports of foreign technology and equipment from the West. Iraq was scheduled to start operations of its first pilot cascade of 100 centrifuges in mid-1993.[23]

Another site involved in the EMIS program was the Al Rabiyah Manufacturing Plant. This plant had large mechanical workshops designed and built for the manufacture of large metal components for the EMIS program. This installation was untouched during the Gulf War but it was attacked by the United States in January 1993 with Tomahawk land-attack missiles (TLAM).[24]

The greatest surprise for Western nuclear experts was the Iraqi success in using EMIS. Using computer controls and fiberoptic links, the Iraqis developed highly automated isotope separators. EMIS is considered the worst of all methods to enrich uranium because they require enormous quantities of electricity to power the large magnets or calutrons. But Iraq has plenty of electrical power from petroleum and hydroelectric sources. This

---

[23]Ebert, "Iraq: Its Nuclear Past As a Way of Assessing Its Nuclear Future," p. 7, and IAEA, "Fact Sheet," p. 4.

[24]IAEA, "Fact Sheet," p. 4, and United Nations, Security Council, *Report on the Seventeenth IAEA On-Site Inspection in Iraq Under Security Council Resolution 687 (1991)* (January 25–31, 1993), U.N. Doc. S/25411 (March 13, 1993), p. 8.

technology has been discarded by the West long ago. Indeed, "calutrons are the Model-Ts of the Nuclear Age."[25]

Experiments on the EMIS system and in uranium recovery and UCl_4 production were conducted at Al Tuwaitha. Manufacturing facilities were centered at (1) Al Ameer— prototype EMIS components, (2) Al Radwan and Al Amir— magnet cores, return irons, ion sources, and collector parts, (3) SEHEE (Daura)—vacuum chamber parts, (4) Salladine and the Dijjla laboratory—assembling of electrical control panels, and (5) Al Tuwaitha—manufacturing of coils. An EMIS facility was operating at Tarmiya and another one was planned at Ash Sharqat; Al Jesira produced the UCl_4 required.[26]

Sites Related to Weaponization. This is a developing story and the IAEA will address weaponization issues further in 1996. For now, we know that the Iraqis had an ongoing weaponization program at five main facilities. Three additional facilities are added in Table 2–6. Efforts to develop high explosive triggers for the bombs took place at Al Qa Qaa, Hatteen, Al Atheer, and Al Hadre. Development of an implosion device was underway at Al Tuwaitha.[27]

While initial efforts at weaponization were begun at Al Tuwaitha, this work was later shifted to Al Atheer, which became

---

[25]Smith and Frankel, "Saddam's Nuclear-Weapons Dream," p. A33. See also Peter D. Zimmerman, "Proliferation: Bronze Medal Technology Is Enough," *Orbis* (Winter 1994), pp. 76–77.

[26]United Nations, Security Council, *Report on the Seventh IAEA On-Site Inspection in Iraq*, p. 22.

[27]*Ibid.*, p. 8.

the primary center for Iraq's weapons design, development and testing efforts. Because this installation was unknown to U.S. planners, it was virtually untouched during the war. The sketches used by IAEA of Al Atheer and the adjacent Hatteen high explosives test site and bunker are shown at Figure 2–6.

Weapon-making involves the design and fabrication of the fissile core, nonnuclear components (chemical explosives, detonator, fuze, neutron inhibitor, reflector), and assembly. High explosives are exploded by detonators timed electronically by a fuzing system. The nuclear chain reaction is triggered by an "initiator" that injects a burst of neutrons into the fissile core at the right moment.

The Iraqis were examining the implosion technique for their nuclear weapons. A shell of chemical high-explosive surrounding the nuclear material is designed to be detonated nearly simultaneously at multiple points to compress the nuclear material rapidly to form a supercritical mass. The implosion technique requires substantially less nuclear material than the gun-assembly method.[28]

Other Sites Supporting Iraq's Nuclear Weapons Program. Table 2–7 lists seventeen facilities providing support for Iraq's nuclear weapons program. A couple of these were storage locations for key components, others had the precision machine tools necessary in the manufacture of key elements, some provided electronic and computer support, and a range of other

---

[28]U.S., Office of Technology Assessment, *Technologies Underlying Weapons of Mass Destruction*, pp. 173–75.

Table 2–6
Sites Related to Weaponization

Name	Also Known As	Location	Purpose
Al Atheer Materials Center		Adjacent to Hatteen HE Test Site	Partially completed nuclear weapons development and production site. Eight specialized process buildings comprising 350,000 sq ft of lab space. Complex for developing and testing high explosive trigger of the bomb. Virtually untouched in Gulf War.
Al Hadre	Al Hatra, Al Madre		High Explosives (HE) Test Site, high explosives research and hydrodynamic studies. Damaged in Gulf War.
Hatteen	Al Hatteen HE Test Site	Al Atheer	HE weaponization and research.
Al Kinde	SAAD-16; Al Mosul	3 mi N of Mosul	Contains equipment which could be modified to serve nuclear HE development needs. Badly damaged in Gulf War.
Al Mansour Electronics Plant			Produced electronic components as part of weaponization feasibility study.

Al Musayyib	Musayyib, Al Mussayib Thermal Power Station	35 mi SW of Baghdad; adjacent to Al Atheer	Materials research and HE test site. Test range for shaped charges, power plant, nuclear weapons laboratories. Production of hydrogen gas.
Al Nafad		1 mi from Al Tuwaitha	Equipment storage area.
Al Qa Qaa		Iskandrariya	High explosives weaponization. Development of non-nuclear components and explosives for nuclear weapons: production and casting for weapon, pressing and machining, main explosive structure of weapon, explosive lens building, and lens assembly.
Al Tuwaitha Nuclear Research Center	Baghdad Nuclear Research Center		Technical headquarters of the Iraqi nuclear program and site of many R&D functions. Approximately 100 structures in the compound. Mostly destroyed in Gulf War.

Figure 2–6
Al Atheer

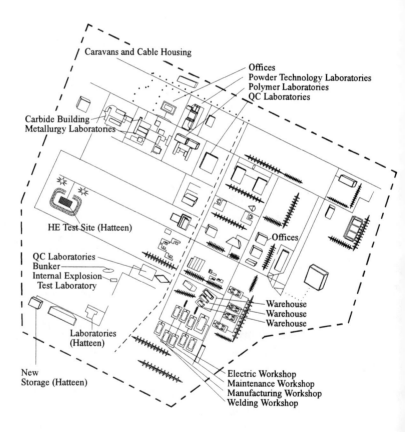

Caravans and Cable Housing

Offices
Powder Technology Laboratories
Polymer Laboratories
QC Laboratories

Carbide Building
Metallurgy Laboratories

HE Test Site (Hatteen)

Offices

QC Laboratories
Bunker
Internal Explosion
Test Laboratory

Warehouse
Warehouse
Warehouse

Laboratories
(Hatteen)

New
Storage (Hatteen)

Electric Workshop
Maintenance Workshop
Manufacturing Workshop
Welding Workshop

technical support activities are included. The data on these facilities gleaned from the IAEA on-site inspection reports tell only a part of the story.

Documentary evidence revealed that two Iraqi facilities had had a program for developing an implosion-type nuclear weapon; other documents linked nuclear weapons to a surface-to-surface missile project. The extensive weaponization program had been centered on Al Tuwaitha and Al Atheer, including work with neutron irradiators and plans for external irradiating high-explosive components, exploding bridge wire detonators and firing sets for multiple detonator systems.[29]

Producing a nuclear weapon would be a formidable task for even a country such as Iraq, which has a large and well-developed electronic, chemical, and metallurgical manufacturing capability. Development of a nuclear weapon, even crude ones such as those used against Hiroshima and Nagasaki, for example, involves thousands of parts. The Los Alamos National Laboratory detailed the components needed to make a nuclear weapon and where to obtain them: the document is 500 pages long.[30] For this reason, a number of other facilities were involved in Iraq's nuclear program that would not at first glance appear to be critical targets. In many cases, however, they contain precision machine tools or computer equipment the Iraqis would find necessary to complete the project, and thus would be legitimate military targets. As

---

[29]United Nations, Security Council, *Report by the Executive Chairman of the Special Commission Established by the Secretary-General Pursuant to Paragraph 9 (b) (i) Security Council Resolution 687 (1991)*, U.N. Doc. S/23165 (October 25, 1991), p. 23.

[30]Albright and Hibbs, "Iraq and the Bomb," pp. 18–19.

Table 2–7
Other Sites Supporting Iraq's Nuclear Weapons Program

Name	Also Known As	Location	Purpose
Abu Gharib	Abu Ghrayb, Khan Dari Warehouse Complex		Military base and fuel rod storage. Receiving point for foreign shipments.
Al Adaya		20 mi S of Al Jesira main site	Burial location for some Al Jesira equipment.
Al Karama			Machine tools subject to routine monitoring.
Al Tuwaitha	Transport/Engineering Service Center	Just outside the Tuwaitha berm	Provided dispatching services for movement between IAEA sites. Provided utility and engineering design services to Tuwaitha and other IAEA sites.
Al Zawra			Support role in nuclear weapons program.
Babyl Company			Machine tools subject to routine monitoring.
Dijila Electronics Plant	Al Zaura, Al Dijla	Proximity of Al Rabiya	Supported electronics fabrication activities for the Iraqi Atomic Energy Commission (IAEC). Not damaged in Gulf War.
Falluja Lead Factory			Recycle old lead acid car batteries. Possible storage area for precision machined graphite plates.
Ibn Al Haytham			Space Research Center. Contains specific pieces of equipment identified in procurement data or in Iraqi declarations.

Iskandariya	State Enterprise for Mechanical Industries; Al Iskandariya		Two factories: car and tractor. Contains machine tools capable of producing equipment for nuclear power.
Latifiyah	Nuclear Agriculture Research Farm	25 mi S of Baghdad	Iraqi Atomic Energy Commission stored equipment and documents at the farm.
MIC (Military Industry Commission) Storage		Immediate proximity of Baghdad North Gate	Former PC-3 Headquarters. Consists of five high-rise buildings.
Mosul Military Production Facilities	SAAD-24		Contains equipment capable of involvement in nuclear program. Heavily damaged in Gulf War. Each building was hit by precision bombing.
National Computer Center		Baghdad	NEC 750 computer mentioned in several PC 3 reports.
Rashdiya	Al Rashidiya Engineering Design Center	Proximity of Baghdad North Bridge	Engineering Design Center. Some R&D activities for centrifuges carried out.
Taji E Glass Plant	Taji Fiberglass Factory		Produces fiberglass and products made from fiberglass. Contains equipment subject to routine monitoring.
Unidentified		Near Al Jesira	Storage of nuclear waste drums and filters.

IAEA inspectors delved deeper into Iraq's nuclear program, they often found references to other locations that could have played a role in the program. Subsequent inspection of these facilities, including a bicycle plant that had received high voltage electronic tubes, an electronics factory, and a prison, revealed that their functions were consistent with their stated uses, and they were not involved in the nuclear program.

The Iraqi Crash Program. The Western response to Iraq's August 1990 takeover of Kuwait triggered an Iraqi "crash program" to obtain a nuclear weapon or device by the end of April 1991. During technical talks between Iraqi and IAEA officials in August 1995, the Iraqis explained that the crash program planned to divert from IAEA safeguards the highly enriched uranium (HEU) material from the fuel of the French- and Russian-supplied research reactors to Iraq's clandestine nuclear weapons program. Among the tasks necessary to make these possible were the reprocessing of both the unirradiated and irradiated fuel to recover HEU and the re-enrichment of the eighty percent enriched material from the research reactors.

The successful implementation of the crash program would have resulted in Iraq extracting some twenty-five kilograms of HEU, with an average enrichment of eighty-six percent from fresh and lightly irradiated research reactor fuel. Being more problematic, they probably could not have extracted HEU from the irradiated research reactor fuel until the end of October 1991. According to the Iraqi accounts, the production of a nuclear weapon from the HEU was dependent on a host of design and weaponization activities that would have made it impossible for them to deploy or test a nuclear weapons or device before the end of 1992.

By January 1991, the Iraqis had completed the design and construction of a small-scale pilot reprocessing plant in the hot cells of Building 22 (LAMA) at Al Tuwaitha and were ready to commence operations. However, the Coalition air strikes on Al Tuwaitha destroyed the building. The Iraqis explained that they removed and destroyed the principal components of the pilot plant to remove evidence of the project.

With Iraq's weaponization program not yet adequately developed, the crash program apparently was premature. According to Iraqi officials, it could have taken them a year and possibly extending into 1992 to produce a single nuclear weapon or device.[31]

Implications for Counterproliferation Operations

With Western intelligence agencies "snookered" by Iraq's deception and persistent boldness in doing the unthinkable (such as using calutrons and moving machinery out of buildings before they could be bombed), U.S. target planners operated in the blind. How could this have happened? One target, at most two, were identified at the beginning of the war and six or seven more had been added by the conflict's end; following the IAEA inspections, the number of nuclear facilities mushroomed to fifty-six, some

---

[31]United Nations, Security Council, *Report on the Twenty-Eighth IAEA On-Site Inspection in Iraq Under Security Council Resolution 687 (1991)*, U.N. Doc. S/1995/1003 (December 1, 1995), pp. 3–17.

seven times greater (See Figure 2–7.) A part of the reason for this wide difference is that the U.S. consistently underestimated Iraq's nuclear capabilities before the Gulf War. According to one newspaper account, the U.S. Department of Energy and Department of State were firm in their views in 1989 and 1990 that the Iraqis lacked the technical competence for rapid breakthroughs to nuclear weapons and that several years of effort would be required. Although the CIA disagreed with this judgment, it did not offer a dissenting view in government estimates. Such faulty estimates prepared the ground for American military commanders and their planners to misperceive Iraq's true nuclear weapons potential.[32]

The debate over when Iraq may have had nuclear weapons continues as additional evidence of its clandestine program comes to light. For instance, Paul Leventhal of the Nuclear Control Institute thinks "the Iraqi nuclear program is far from dead." He points out that when the Iraqis claim they were only three months away from a nuclear weapons, they must have had the components ready to assemble the device once they had sufficient fissile material. Yet, the IAEA did not find the components after the war. If the Iraqis have the essential parts of a nuclear device hidden, does it mean they are still three months away from a nuclear explosive? Some believe the Iraqis may be shopping for fissile material on the international black market as well as for the nuclear expertise necessary to hurdle technical difficulties.[33]

---

[32]R. Jeffrey Smith, "Iraq's Nuclear Prowess Underestimated by U.S.," *Washington Post* (October 13, 1991), p. A45.

[33]As quoted in Barbara Crossette, "Experts Doubt Iraq's Claims On A-Bomb," *New York Times* (August 30, 1995), p. A6.

Figure 2–7
Fifty-Six Iraqi Nuclear Facilities

Baghdad Facilities
Al Shaheed Brass Factory
Babyl Company
Geological Survey Institute
Military Industrial Commission Storage
Nahrawan
Nassar Works
Nassr State Establishment
National Computer Center
Petrochemical Projects
Saddam Works
Taji Center for Metallurgical Industries
Taji E Glass Plant

Al Tuwaitha Facilities
Al Hamath
Al Nafad
Ash Shakyili
Garf Al Naddaf

Other Facilities
Abu Gharib
Al Karama
Al Mansour
Al Zawra
Ibn Al Haytham

Not until November 1995 did it come to light that the Iraqis were also working on a radiological weapon, a device that scatters deadly radioactive material without causing a nuclear explosion.[34] The Iraqis already had demonstrated their ability to tap uranium in civilian reactors by separating three grams of plutonium from irradiated reactor fuel in violation of safeguard agreements with the IAEA. It remains unclear as to what material the Iraqis would have to have irradiated through their research and development reactors to produce effective radioactive weapons.[35]

The Western countries were very fortunate that Saddam Hussein acted before a few nuclear weapons were produced and readied for delivery. The Iraqi nuclear weapons program illustrates some alternative pathways that nuclear proliferators may take in the future. While the Iraqi case provides examples about what to look for in the future along different proliferation pathways, tomorrow's proliferators are also informed by the Gulf War bombing and post-war IAEA inspections of what the telltale signatures of a nuclear program might be—they are likely to make our detection of their nuclear weapons programs far more difficult than was the case in Iraq.

The Iraqis demonstrated that plutonium is not necessary to produce nuclear weapons—highly enriched uranium is "good enough." Moreover, nuclear expertise is readily available and reportedly some nuclear scientists and engineers from the former Soviet Union are willing to cross borders if the price is right.

---

[34]"U.N. Official: Iraq Worked On Radiological Arms," *Washington Post* (November 8, 1995), p. A25.

[35]United Nations, Security Council, *Report of the Executive Chairman of the Special Commission*, U.N. Doc. S/23165, pp. 22–23.

Much of the equipment necessary to produce nuclear weapons falls into the dual-use, civilian and military category, opening the door to plausible denials of the actual end uses. The international black market of fissile material from countries of the former Soviet Union may provide viable alternatives to indigenous production of HEU. Moreover, the end of the Cold War has provided a political rationale in some countries to pursue nuclear weapons programs.

The U.S. armed forces need to begin preparing today for possible counter-nuclear operations against countries holding a few nuclear weapons in the years ahead. Such a counterproliferation mission will require exquisite intelligence, rigorous planning, and the weapons, supporting systems, and delivery platforms necessary to assure destruction of the nuclear targets when authorized by the President. The Iraqi case, both in terms of pre-war intelligence and post-war on-site inspections, demonstrates that there will *always* be great uncertainties about nuclear weapons programs. The United States will need military forces that reflect flexibility and versatility in the conduct of future counter-nuclear operations.

Chapter 3
Shading a Biological Weapons Program

Biological weapons contain infectious or toxic agents that cause disease or death.[1] Derived from natural sources such as anthrax and botulism, these weapons can inflict massive casualties. Botulism toxin, for instance, is about three million times more potent than the nerve agent, sarin, used in chemical weapons. A Scud missile warhead filled with botulinum could contaminate an area of 3,700 square kilometers, or sixteen times greater, than a warhead filled with sarin could. Once symptoms occur, treatment offers little hope; death can follow in as few as twelve hours. Anthrax is more persistent and can cover a much larger area than botulinum using the same delivery means; following exposure, anthrax takes two to four days to kill. Fifty kilograms (a little over 100 pounds) of dried anthrax, when used against a city in a suitable aerosol form, can cover more than twenty square kilometers and, according to the World Health Organization, inflict "tens to hundreds of thousands of deaths." The means to detect biological warfare agents rapidly do not exist.[2]

---

[1]"Biological agents are infectious microorganisms that reproduce within the host to cause incapacitating or fatal illness. Toxins, being poisonous chemicals manufactured by living organisms, have characteristics of both chemical and biological agents." See U.S., Congress, Office of Technology Assessment, *Technologies Underlying Weapons of Mass Destruction*, p. 73.

[2]U.S., Department of Defense, *Conduct of the Persian Gulf War*, p. 15, and the World Health Organization as quoted in Michael Eisenstadt, *Like a Phoenix From the Ashes? The Future of Iraqi Military Power*, Policy Paper No. 36 (Washington, D.C.: Washington Institute for Near East Policy, 1993), p. 33.

After successfully hiding the extensive scope of its biological weapons program from Gulf War targeters and United Nations inspectors after the war, Iraq, according to its latest declaration in August 1995, admitted to have weaponized 166 bombs and twenty-five Scud warheads with biological warfare agents just before the Coalition attacks were launched in January 1991.[3] If these botulinum- and anthrax-filled weapons had been used, they could have killed hundreds of thousands of people and inflicted diseases on the cities and military bases in Saudi Arabia, Israel, and other countries, as well as aircraft carriers, amphibious forces afloat, and other sea-based units within reach of its aircraft and missiles. The Iraqi leaders say that they did not use the biological weapons for fear that the United States would retaliate with nuclear weapons.[4]

The United States and its Coalition partners may have "dodged a bullet" in this case, but there is little to offer confidence for the future. If the Coalition had pursued Saddam Hussein and his remaining military forces into Baghdad, might not the Iraqi president, in a dying sting of his regime, have unleashed his botulinum and anthrax weapons against Riyadh, Tel Aviv, military bases, and naval ships at sea within reach of his aircraft and missiles?

---

[3]United Nations, Security Council, *Report of the Secretary-General on the Status of the Implementation of the Special Commission's Plan for the Ongoing Monitoring and Verification of Iraq's Compliance with Relevant Parts of Section C of Security Council Resolution 687 (1991)*, U.N. Doc. S/1995/864 (October 11, 1995), p. 8.

[4]R. Jeffrey Smith, "U.N. Says Iraqis Prepared Germ Weapons in Gulf War," *Washington Post* (August 26, 1995), pp. A1, A19.

Despite the 125 signatories to the 1975 Bacteriological (Biological) and Toxin Weapons Convention, the development of these weapons continues to increase. Biological weapons are especially difficult to control because the raw materials and equipment used have many legitimate civilian uses, particularly in the pharmaceutical, medical and food industries. In 1972, when the Convention was first opened for signature, four countries were suspected of having biological weapons programs—the number rose to ten countries in 1992, some of which are members of the Convention. The United States has expanded and refined its export controls pertaining to biological organisms, toxins, and related equipment; other countries (the Australia Group) have followed suit. Yet, the items necessary to create and sustain a biological weapons program are readily available on the world market.[5]

Biological weapons continue to be developed as "the poor man's atomic bomb" by a growing number of countries. The ability to differentiate between offensive biological weapons and defensive measures such as vaccines does not exist. The same laboratory techniques and expertise are necessary for both offensive and defensive programs. Militarily significant quantities of pathogens and toxins, as demonstrated in the Iraqi case, can be produced through advanced biotechnology in easily concealable facilities.

About thirty out of hundreds of pathogenic microbes that indirectly or directly affect humans are considered as likely

---

[5]U.S., General Accounting Office, *Arms Control: U.S. and International Efforts to Ban Biological Weapons*, GAO/NSIAD-93-113 (December 1992), pp. 2–3.

warfare agents. Some of the more common biological agents and toxins are summarized in Table 3–1.[6]

What lessons for the future can we obtain from the Gulf War's bombing campaign, the U.N.'s on-site inspections after the war, and Iraq's subsequent admission that it had produced and weaponized biological weapons? The most likely scenarios involving U.S. forces in future contingencies are also the most lethal ones. Adversaries will possess biological weapons; some will be declared, and others undeclared. How can the U.S. conduct effective military operations that will best achieve its objectives in the face of biological threats to American forces on land and at sea?

The Persian Gulf War

Eliminating Iraq's presumed biological weapons held a high priority for the Coalition's planners and commanders as being critical to minimizing friendly casualties. Intelligence sources were clear that Iraq's biological weapons program began sometime in the late 1970s, but the exact nature and current status of the program were tenuous. As a consequence, planners considered actual biological weapons use as being remote. Nonetheless, this target category was included in the weapons of mass destruction cluster along with nuclear and chemical targets that made up one of the Iraqi centers of gravity identified for priority bombing. Some 970 strikes pounded the Iraqis nuclear,

---

[6]U.S., Congress, Office of Technology Assessment, *Technologies Underlying Weapons of Mass Destruction*, pp. 71–76.

Table 3–1
Some Common Biological and Toxin Agents

Potential Warfare Agent	Agent Characteristics and Examples
Pathogens:	
Bacteria	Single-cell organisms that are causative agents of anthrax, brucellosis, tularemia, plague, and other diseases.
Rickettsiae	Microorganisms that resemble bacteria but differ in that they are intracellular parasites that can reproduce only inside animal cells, e.g., typhus, Rocky Mountain Spotted Fever, Q Fever.
Viruses	Intracellular parasites that are 100 times smaller than bacteria, e.g., hemorrhagic fever such as Ebola or Lassa Fever.
Toxins:	
Protein Toxins	Most protein toxins are large proteins, e.g., various strains of staphylococcus aureus, a major bacterial pathogen, botulinum (the most poisonous substance known), cholera, tetanus, diphtheria.
Nonprotein Toxins	Small organic molecules that often have a complex chemical structure; high toxicity, no antidotes, heat stability, speed of action, e.g., tetrodotoxin (produced by a puffer fish), palytoxin (made by a soft red Hawaiian coral), batrachotoxin (secreted by poisonous frogs indigenous to western Columbia).

Source: U.S., Congress, Office of Technology Assessment, *Technologies Underlying Weapons of Mass Destruction* (Washington, D.C.: Government Printing Office, 1993), pp. 76–81.

biological, and chemical facilities; the bulk of the attacks occurred against what was considered as the most likely threat—chemical weapons.[7]

The Coalition was successful in identifying suspected research and production facilities at Salman Pak, Taji, Al Latifiyah, and two near Abu Ghurayb. Iraq also had scattered throughout the country twenty-one specially designed, refrigerated bunkers that were suspected of containing biological or other special weapons. Eighteen of these had been identified before the war. It was believed that Iraq had produced anthrax spores and botulinum toxin as agents and that it was pursuing other agents as well. Baghdad's lack of operational experience with biological weapons made U.S. intelligence estimates very tenuous. Analysts "could only speculate that Saddam might resort to biological weapons to preempt a coalition offensive, achieve certain battlefield objectives, or save himself and his regime from destruction."[8]

The biological targets also created new operational problems for planners. Given the virulent microorganisms and toxins that might be widely dispersed as a result of bombing, "they feared high-explosive bombs striking these buildings might produce clouds of aerosolized agents that could travel long distances, contaminate large areas, and posing thousands or even millions

---

[7]U.S., Department of the Air Force, *Gulf War Air Power Survey: Operations and Effects and Effectiveness*, Vol II, Part 2, p. 322.

[8]U.S., Department of the Air Force, *Gulf War Air Power Survey: Planning and Command and Control*, Vol. I, Part 1, pp. 160–61.

of people within and outside Iraq."[9] Lieutenant General Charles Horner, commander of the Air Force component assigned to the U.S. Central Command, believed that the risk of bombing the biological bunkers could be managed, and he argued that "if there was a small outbreak of disease as result, that would be a lesson to third world nations about the risks of producing biological weapons."[10] Such attacks were planned carefully to minimize the chances of biological contamination. Secondary explosions occurred when some of the bunkers were hit, indicating that they contained weapons of some sort.

Before the Gulf War only one known biological target (Salman Pak) was present on U.S. targets lists, plus eighteen suspected storage bunkers. By the war's end, these numbers had grown to five and twenty-one respectively. All of the infrastructure targets were damaged in the air attacks and nineteen of the suspected biological storage bunkers were destroyed (The last two were discovered too late to strike before the war's end.). One of the infrastructure targets, the Al Kinde veterinary company, one of the two suspected production plants in Abu Ghurayb, was funded by a grant from the U.N. Food and Agriculture Organization, which after the war U.N. inspectors found no evidence of prohibited activity.[11] In the end, the Coalition planners were unsure whether any biological weapons had been in any of the bombed facilities. In February 1991, however, an article in the Egyptian press which was later picked up in Soviet publications, reported that an attack on a biological

---

[9]*Ibid.*

[10]As quoted in Gordon and Trainor, *The Generals' War*, p. 192.

[11]*Ibid.*, pp. 182–83.

facility outside of Baghdad had resulted in the deaths of fifty guards from a rapidly progressing disease that spread into Baghdad. No other mention of the alleged incident has been made.[12]

Iraq's Biological Weapons Program

U.N. Inspections. The first biological weapons inspection by the U.N. Special Commission was carried out in August 1991. The inspectors conducted a full and detailed inspection of Salman Pak. The Iraqis claimed that a biological research program had been carried out between 1986 and 1990, at which point all stocks were destroyed. The microorganisms included botulinum toxin, perfringens (gas gangrene), and Bacillus anthracis (anthrax). Further evidence showed that the Iraqis had possessed other biological warfare agents, including Brucellas abortus, Brucella melitensis, Francisella tularensis, and various strains of botulinum. No biological weapons or evidence of weaponization were found. The second U.N. inspection included ten declared and undeclared sites (including a pharmaceutical plant, a blood bank, vaccine production facilities), four without advance notice. Neither biological weapons nor facilities for filling weapons were found.[13]

While acknowledging that it had acquired imports of complex growth media essential for hospital diagnostic laboratories, Iraq

---

[12]U.S., Department of the Air Force, *Gulf War Air Power Survey: Operations and Effects and Effectiveness*, Vol. II, Part 1, pp. 229–30.

[13]United Nations, Security Council, *Report By the Executive Chairman of the Special Commission*, U.N. Doc. S/23165 (October 25, 1991), p. 30.

denied that any of them had been diverted to biological warfare purposes. Several incongruities marred the Iraqi explanations. Total hospital consumption of the imported growth media (for preparing nutrient solutions for bacteria cultures) was less than 200 kilograms per year for the entire country. Yet, in 1988 alone Iraq imported 39,000 kilograms of a medium suitable also for the production of anthrax and botulinum. The U.N. Special Commission could account only for twenty-two tons of the thirty-nine tons imported in 1988. Iraqi explanations were unsatisfactory also concerning the acquisition of dual-purpose biological equipment and supplies. Four filling machines at Salman Pak, ostensibly for a biopesticide project, also could be used for filling bacterial warfare agent into munitions or containers. A spray dryer procured in 1989 was claimed to be for the biopesticide project, but had the capability of producing dry matter with particle sizes associated with the efficient dispersion of biological warfare agents, not with the production of biopesticides.

The construction of biological facilities at Al Hakam raised other inconsistencies in the Iraqi argument. Iraq claimed the facility was intended only as a single-cell protein plant for the production of animal feed. The original design for Al Hakam, however, included many costly features associated with production of toxic or infectious materials. A sophisticated air filtration system and security arrangements consistent with a military facility or one producing toxic or pathogenic material were inconsistent with a commercial production facility. The facility is remote, heavily guarded, and contains berms for blast protection and modern disposal buildings. Furthermore, Iraq falsified information on an end-user certificate for a fermenter at Al Hakam and for spare parts for equipment located there.

By early 1995, the U.N. Special Commission had been frustrated in its attempts to establish a baseline of Iraq's dual-purpose biological capabilities. Iraq submitted incomplete and inaccurate initial declarations, numerous inconsistencies were apparent between Iraq's declarations and the findings of the inspection teams, and undeclared movement of items often resulted in further inconsistencies with the findings of the inspection teams.   "All this made it impossible for the Commission to establish firm baseline data from which to start its monitoring of Iraq's biological activities."[14] Nonetheless, it was quite clear that Iraq had obtained all of the items and materials required to produce biological warfare agents. While the U.N. inspectors could not find a "smoking gun" to serve as irrefutable proof of the Iraqis weaponization of biological agents, they had uncovered a sufficient number of concerns to prevent Iraq from receiving a clean bill of health from the United Nations Security Council. Specifically, the media used for nutrient solutions, the overbuilding at Al Hakam, and unexplained weaponization activities added up to the list of serious unanswered questions.

A Locked Chicken House. After more than four years of stonewalling and misleading the U.N. inspectors, the Iraqi biological weapons program was quickly unraveled to show new

---

[14]United Nations, Security Council, *Report of the Secretary-General on the Status of the Implementation of the Special Commission's Plan For the Ongoing Monitoring and Verification of Iraq's Compliance With Relevant Parts of Section C of Security Council Resolution 687 (1991)*, U.N. Doc. S/1995/284 April 10, 1995), pp. 18–20.

insights and raise yet another round of unanswered questions. Rolf Ekeus, Executive Chairman of the U.N. Special Commission on Iraq, received an oral admission from senior Iraqi officials while visiting Baghdad on July 1, 1995, that Iraq had produced a number of biological agents. The Iraqis denied, however, having weaponized the agents. On July 17, Saddam Hussein in a speech in Baghdad indicated that his government would cease cooperation with the Security Council if there was no progress on the lifting of sanctions and the oil embargo. On August 4, Mr. Ekeus again arrived in Baghdad for discussions; the Iraqis claimed to have given him full, final and complete disclosure of their biological weapons programs, insisting that no weaponization had occurred. On August 7, Lieutenant General Hussein Kamel, director of the Ministry of Industry and Military Industrialization defected to Jordan. General Kamel had been Saddam Hussein's number-two man in the regime and the president's son-in-law. He was joined in Jordan by about thirty other officers and their families, including the chief of Saddam Hussein's personal bodyguard.[15]

Mr. Ekeus returned to Baghdad on August 17 upon invitation by the Iraqi Government. Iraqi officials stated that General Kamel had hidden information on prohibited programs from the leadership, which they could now disclose. The Iraqis admitted to having had a much larger biological weapons program than

---

[15]Lt. Gen. Hussein Kamel and his brother, Colonel Saddam Kamel, who were both sons-in-law of Saddam Hussein, received permission in early February 1996 to return to Iraq. Once they were inside the country, however, both men were gunned-down on February 23, 1996, by members of their own family. See "King Hussein Denounces Killing of Repatriates," *Washington Post* (February 26, 1996), p. A20.

they had previously disclosed and that some weaponization had occurred just before the outbreak of the Gulf War. As Mr. Ekeus and his party prepared for departure, General Amer Rashid Ubeidi contacted the chairman and suggested that on his way to the airfield for departure he stop off at a farm that had belonged to General Hussein Kamel. During their visit at the farm, the chairman and his party were shown numerous containers with miscellaneous equipment in them. When they opened a locked chicken house, the U.N. party discovered about 150 wooden and metal boxes packed with documentation on all biological programs, including microfiches, computer diskettes, videotapes, photographs, and prohibited hardware components.   The Commission's personnel transferred this material to its Baghdad Monitoring and Verification Center.

The six major sites associated with Iraq's biological warfare program are depicted in Figure 3–1 and details on these and other sites are presented in Table 3–2. Iraq tested biological aerial bombs as early as 1988 at Al Muthanna. Live firings of 122 mm rockets with botulinum toxin and aflatoxin were conducted in May 1990. Following Iraq's August 2, 1990, invasion of Iraq, its biological weapons warfare program shifted into high gear. The foot and mouth disease plant at Daura was converted to biological weapons production. By January 1991, the plant had produced at least 5,400 liters of concentrated botulinum toxin. Meanwhile, Al Hakam turned to production of anthrax. Iraq declared it also had produced 340 liters of concentrated perfringens by the time the Gulf War broke out.

Figure 3–1
Seven Iraqi Biological Weapons Facilities

Table 3–2

Iraq's Biological Warfare Facilities

Name	Also Known As	Location	Purpose
Al Hakam	Al Hakem, Al Hakim	37 mi SW of Baghdad	A seven-square-mile complex designed and constructed as Iraq's main biological agent production facility; produced thousands of liters of anthrax and botulinum toxin, hundreds of liters of perfringen (gas gangrene); not damaged in the Gulf War.
Baghdad Biological Research Center	Scientific Research Council	Baghdad	Main site for much biological weapons research.
Daura Foot and Mouth Disease Vaccine Facility		SE outskirts of Baghdad	Site of initial research on viral warfare agents including hemorrhagic conjunctivitis, human rotavirus, and camelpox; produced thousands of liters of botulinum toxin.
Al Fudaliyah		NE outskirts of Baghdad	Agricultural and Water Resources Research Center; produced 2,000 liters of aflatoxin.

Al Muthanna State Establishment	75 mi NW of Baghdad	Drawing on its chemical weapons munitions technology, weaponization expertise given to the biological warfare program; Al Muthanna was heavily damaged in the Gulf War.
Salman Pak	25 mi SE of Baghdad	Laboratory scale research on anthrax, botulinum toxin, perfringens (gas gangrene), mycotoxins, aflatoxins, and ricin; conducted initial scale-up production research; heavily damaged in the Gulf War.
Taji Single Cell Protein Plant	6 mi NW of Baghdad	Converted for the production of hundreds of liters of botulinum toxin.
Storage Bunkers		19 of 21 refrigerated storage facilities suspected of storing biological weapons were struck, including four at Salman Pak
Sites Bombed But not Connected With Biological Warfare		Abu Ghuayb, the so-called baby milk factory, and Al Latifiyah appeared to U.N. inspectors after the war to have not been involved.
79 U.N. Monitoring Sites		Facilities with dual-purpose equipment and activities; civil functions that could speedily convert to biological warfare purposes

A program was initiated in December 1990 to develop a biological spray tank to spew up to 2,000 liters of anthrax over a target from an aircraft or a remotely piloted vehicle. Field tests were considered a failure but three additional spray tanks were prepared and stored. In December 1990, bombs were chosen as the munition for aerial delivery: 100 were filled with botulinum toxin, fifty with anthrax, and sixteen with aflatoxin, a cancer-causing agent. Twenty-five Scud warheads also were filled: thirteen with botulinum toxin, ten with anthrax, and two with aflatoxin. These weapons were stored at four locations throughout the war.

The U.N. Commission, despite Iraq's disclosures in August 1995, still does not believe that a full and correct accounting of the biological weapons program has been given. In its tireless quest for a complete disclosure of Iraq's proscribed biological weapons program, the Commission, as shown in Table 3–3, continues to monitor seventy-nine sites throughout the country.[16]

In the event of war, these seventy-nine sites logically would be listed as "suspected" biological weapons targets by intelligence and operational planners. The number of locations points to the difficulty of destroying a priority target set when extensive dual-use materials and equipment are involved. During the Gulf War, the Coalition air forces struck Al Kinde, which was dedicated to

---

[16]United Nations, Security Council, *Report of the Secretary-General*, U.N. Doc. S/1995/864, pp. 1–29; "Saddam Hopes BW Confession Is Enough To Convince USA," *Jane's Defence Weekly* (September 2, 1995), p. 27; and "Iraq Finally Admits Building Biological Weapon Arsenal," *Jane's Defence Weekly* (July 15, 1995), p. 3.

civilian production, but never hit Al Hakam, a producer of the deadliest biological agents known to man.

Implications for Counterproliferation Operations

A French microbiologist who participated in two of the U.N. inspections told U.S. researchers from the General Accounting Office that he did not believe Iraq had an offensive biological warfare capability. He pointed out that Iraq's fermenters were used for the production of vaccine and were not suitable for

Table 3–3
Biological Monitoring Sites

Sites Monitored	Number
Sites known to have played significant role in the past biological weapons program	5
Vaccine or pharmaceutical facilities	5
Research and university sites which have significant technology or equipment	35
Breweries, distilleries and dairies with dual-purpose capabilities	13
Diagnostic laboratories	8
Acquisition and distribution sites of biological supplies and equipment	5
Facilities associated with biological equipment development	4
Product development organizations	4
Total	**79**

Source: United Nations, Security Council, *Report of the Secretary-General On the Status of the Implementation of Iraq's Compliance With Relevant Parts of Section C of Security Council Resolution 687 (1991)*, U.N. Doc. S/1995/864 (October 11, 1995), pp. 19–20.

producing biological agents. He added that while Iraq's refrigerated storage could be used for biological weapons, they actually were used for meat storage.[17]

Iraq's main production plant for botulinum, the most poisonous substance known, at Al Hakam was never identified by Western intelligence. Air attacks were carried out instead against the Al Kinde veterinary company in Abu Ghurayb and Latifiyah, neither of which was a part of the biological warfare program. Even the twenty-one storage bunkers identified as possible storehouses for biological weapons appear to have been nominated as targets on little more than guesses rather than hard evidence. This should not be surprising, since there were more than 3,000 storage structures in Iraq. By limiting the target set to bunkers, planners would have approximately 800 targets. Since the presence of air conditioning units at some bunkers could be a telltale sign of biological weapons, these bunkers could be nominated for high priority destruction. On the other hand, they simply could be used to store meat, as suggested by the U.N. inspector from France.[18]

The point is that neither Western intelligence nor U.N. inspectors after the war had a clue of the scope and magnitude of the Iraqi biological warfare program. Through use of clever deception measures, hiding equipment, lies, and use of equipment that could be plausibly used for non-military purposes, the Iraqis

---

[17]U.S. General Accounting Office, *Arms Control: U.S. and International Efforts to Ban Biological Weapons*, p. 56.

[18]U.S., Department of the Air Force, *Gulf War Air Power Survey: Operations and Effects and Effectiveness*, Vol. II, Part 1, pp. 230–31, and Gordon and Trainor, *The Generals' War*, p. 182.

were able to spin an image of a country without an offensive biological warfare program and certainly one that did not weaponize munitions and warheads with bacteria and toxins.

The ambiguity and elusiveness of the biological warfare program in Iraq spell trouble for the future. Iraq's own 1995 declarations indicate 19,000 liters of concentrated botulinum toxin with nearly 10,000 filled into munitions, 8,500 liters of concentrated anthrax with some 6,000 liters filled into munitions, and 2,200 liters of concentrated aflatoxin of which 1,580 liters were filled into munitions.[19] The botulinum alone is theoretically enough to kill *fifteen billion* people. Actual use, of course, would confront dissemination inefficiencies but millions still could be killed.[20] The enormity of this kind of killing power in the hands of a person like Saddam Hussein makes one wonder of the consequences if Coalition forces had pursued the Iraqi leader and his elite units into the streets of Baghdad. Even worse is the thought that Western actions could have triggered biological weapons use without even knowing of their existence.

The elusiveness and ambiguity of the biological warfare target set also are reflected by the large number of sites in Iraq (seventy-nine) being monitored by the U.N. Special Commission. This monitoring is a result of the dual-purpose items and activities at these sites and the ease with which civilian facilities can be converted to biological weapons purposes. For the military planner, these dual-purpose sites should be included in the

---

[19]United Nations, Security Council, *Report of the Secretary-General*, S/1995/864, p. 27.

[20]R. Jeffrey Smith, "Iraq Had Program For Germ Warfare," *Washington Post* (July 6, 1995), pp. A1, A17.

biological warfare target set, perhaps labelling them "suspected" sites and awaiting intelligence confirmation before striking them. On the other hand, if U.S. power projection forces are being moved into the theater on land and at sea, the commander may choose to reduce the risks of biological attacks by striking all known and suspected facilities in the biological target set.

Iraq presented U.S. planners a worst case scenario in dealing with a proliferator of biological weapons. Whether by great skill and cunning or by divine intervention, the Coalition avoided triggering Iraq's use of its deadly arsenal. We may not be so fortunate the next time. The United States needs to develop the military plans and forces that will allow timely destruction of an enemy's biological warfare capabilities in a way that prevents dispersal of hazardous materials.

Chapter 4
Expanding a Chemical Weapons Program

In the eyes of Iraqi and other government leaders in the Middle East, chemical weapons represent a cheap and potentially effective response to the conventional threat posed by the modern and sophisticated high-technology battlefield systems of the type that the United States brought to bear during the Gulf War. They also pose a threat of surprise chemical attacks against unprotected enemy civilian populations, attacks that could produce appalling results. Thus, chemical weapons in this region have tactical and strategic purposes and influence national security policies. Since chemical weapons are really truly effective tactically and strategically only when there is an element of surprise in their delivery against unprotected targets, weighty questions are raised for U.S. policymakers over when to attack these weapons, how to mitigate collateral dispersal of chemical agents, and what the best methods are for countering potential terrorist use of these weapons. At the same time, the danger is growing that chemical weapons will become so common in the arsenals of Third World countries that a general acceptance of their use in warfare might grow over time. Such an evolution would have painful consequences for international stability.

Saddam Hussein had made considerable progress in integrating chemical weapons into his conventional force during Iraq's conflict with Iran, 1984–88. According to Iranian figures, the Iraqi chemical attacks inflicted about 50,000 casualties, including some 5,000 deaths.[1] By 1988, the Iraqis had developed an effective offensive doctrine for the use of nerve agents when they are integrated into fire support plans. Both nerve and blister agents were targeted against Iranian command and control facilities, artillery positions, and logistic areas. In fact, the use of these chemical agents may have tipped the scales against Iran and caused its defeat. In the spring of 1988, the Iraqi leader used chemical weapons against Kurdish and Iranian insurgents in the town of Halabjah. Thousands of men, women, and children died.

Iraq had the largest chemical agent production capability in the Third World when it entered Kuwait in 1990. Each year it produced thousands of tons of mustard gas, and nerve agents sarin (GB) and GF. A nonpersistent agent, sarin is relatively easy to produce from readily available precursors. GF is a semipersistent nerve agent similar to soman (GD) that was produced by Iraq after Western nations began to restrict shipments of the precursors for soman.[2] Iraqi chemical agents were to be delivered by aircraft

---

[1]U.S., Congress, Office of Technology Assessment, *Technologies Underlying Weapons of Mass Destruction*, p. 16.

[2]**Mustard gas**, sulfur mustard, is the primary blistering agent used in warfare. It is fairly persistent in the environment and is readily absorbed by the skin and most clothing. Easy to produce compared to the more complex nerve agents, mustard gas can be stored in bulk form or in munitions for decades. It produces painful skin blistering and eye and lung irritations; American and British mustard gas casualties suffered a two to three percent death rate in World War I; the Iranians suffered a similar rate in the 1980s. **Nerve agents** are highly toxic agents that produce convulsions and rapid death by inactivating an enzyme (acetylcholinesterase) that is essential in the normal transmission of nerve impulses. The two classes of nerve agents are designated G and V. The G-series nerve agents are known as tabun (GA), sarin (GB), GC, soman (GD), GE, and GF. These G agents produce casualties

spray tanks as well as rockets, bombs, mortars, artillery shells, and missile warheads.[3]

Unlike the nuclear and biological weapons programs, the Iraqi chemical weapons effort was difficult to hide. Its existence was well known and the chemical agent production had been turned into a nationwide industry. Al Muthanna, the State Establishment for Pesticide Production, directed the national effort. This large complex produced mustard gas and tabun, plus precursors of sarin. Bombing this facility should have been easy pickings for the Coalition air forces. Yet, the Iraqis, anticipating the massive bombing to come, moved equipment used to fill chemical munitions from at least one of three key buildings at Al Muthanna to locations around the country. In all, the Iraqis dispersed their chemical weapons to some twenty locations.[4]

Iraq had built a quite sophisticated capability to manufacture precursor chemicals from simple compounds whose export was not controlled. This so-called "back-integrating" strategy to chemical weapons production was adopted by the Iraqis during the Iran-Iraq War when the Australia Group tried to prevent Iraq from obtaining a key precursor agent of the nerve agent tabun (phosphorus oxychloride). In response, the Iraqis simply built a

---

rapidly by inhalation and by penetrating the skin or eyes at high doses. The V-series nerve agents include VE, VM, and VX. VX is an oily liquid that may persist for weeks or longer in the environment. Ten milligrams of VX on the base skin can kill a medium-sized man. See U.S., Congress, Office of Technology Assessment, *Technologies Underlying Weapons of Mass Destruction*, pp. 21–24.

[3]U.S., Department of Defense, *Conduct of the Persian Gulf War*, p. 15.

[4]Gordon and Trainor, *The Generals' War*, p. 183.

program to be self-sufficient with their own plant to manufacture the precursor, using raw phosphate ore from its phosphate mine at Akashat. Similarly, Baghdad built a large production line at its Basra petrochemicals complex that could produce in excess of 100,000 tons of ethylene per year—thiodiglycol, the immediate precursor of sulfur mustard, can be produced in a batch process by reacting ethylene oxide and hydrogen sulfide. These ingredients are readily available as derivatives of oil and natural gas.

By the time of its invasion of Kuwait, Iraq was well along the way to producing indigenously all of the essential precursors of mustard, tabun, sarin, and VX. In 1988, Baghdad contracted with two West German firms to build three chemical plants at Al Fallujah, about sixty miles west of the capital. These plants were designed to produce a key sarin precursor, phosphorus trichloride, by converting elemental phosphorus and chlorine. The Iraqis also planned to produce another essential ingredient for sarin production, hydrogen fluoride, by using sulfuric acid to extract it from phosphate ore.[5]

The Persian Gulf War

The destruction of Iraqs' nuclear, biological and chemical capabilities was an explicit military objective assigned to Coalition target planners. Despite their obvious differences in production, storage and potential uses, the U.S. intelligence

---

[5]U.S., Congress, Office of Technology Assessment, *Technologies Underlying Weapons of Mass Destruction*, p. 31.

community lumped these three weapons of mass destruction into a single category for analysis and targeting. The blurring of the distinctions between these weapons also clouded estimates of the number of targets in each category. While intelligence authorities agreed in the final months of 1990 that the Iraqis had extensive nuclear, biological, and chemical warfare capabilities, they in fact had few clues about the scope of the nuclear and biological weapons programs. On the chemical warfare estimates, however, they were on much firmer ground.

Coalition planners estimated, for instance, that Iraq produced some 1,000 tons of chemical agents annually at the Samarra and Al Fallujah facilities, including mustard-type blistering agents and tabun and sarin nerve agents. While little doubt existed as to Iraq's chemical warfare capabilities, "what proved harder to predict with certainty was whether, and under what circumstances, Iraq would employ its chemical weapons against the Coalition."[6]

The commanders of the U.S. Central Command and its component commands assumed from the beginning that Iraq would use chemical weapons. President George Bush, Secretary of Defense Dick Cheney, and Chairman of the Joint Chiefs of Staff General Colin Powell were all told by General Schwartzkopf and his commanders that "Iraqi forces will use chemical weapons" in the event of war. Hence, planners were told "to destroy Iraqi capability to produce and [deliver] weapons

---

[6]U.S., Department of the Air Force, *Gulf War Air Power Survey: Summary Report*, p. 78.

of mass destruction" and to do so "as early as possible."[7] Since chemical weapons were considered the most probable threat, this was the exclusive focus of the target set defined in August 1990. Later, nuclear and biological targets were added, broadening the target set. In the meantime, the target list had grown from eight targets in August 1990 to twenty-five targets by December 1990, and leveled off to twenty-three targets by mid-January 1991. The percentage of targets in the nuclear-biological-chemical (NBC) category remained constant at about ten percent of the overall target planning.[8]

The bulk of the 970 strikes against NBC targets was against the Iraqi chemical warfare capabilities. Three primary targets were the three chemical precursor production facilities near Al Fallujah, research centers at Salman Pak (also associated with production of biological warfare toxins), and the chemical munitions production facilities at Samarra. Also listed for attack were suspected storage bunkers for chemical weapons throughout the country. Some of these bunkers, seen mostly at airfields, were "S"-shaped and had unique signatures. Since Coalition intelligence indicated that Iraqi chemical warfare units may have been operating out of Kuwaiti airfields, all hardened shelters on those bases became suspected storage facilities for chemical munitions. A total of thirty suspected chemical storage bunkers were identified throughout the country; air strikes targeted twenty-three and destroyed seventeen. Yet, eight chemical storage bunkers at Samarra escaped destruction since the principal

---

[7]*Ibid.*, p. 161.

[8]*Ibid.*, p. 185.

focus was given to destroying production facilities and aircraft restriking continued to hit main factory elements.   Intelligence identified eight of the thirty chemical storage bunkers in northern Iraq near the cities of Kirkuk and Qayyara. Since U.S. aircraft based in Turkey lacked the requisite capability for delivery of precision-guided weapons, they scattered mines around the bunkers to limit access. On February 11, 1991, ten F-117s ranged north to destroy most of those bunkers.[9]

The Air Force's judgment of the effectiveness of the air campaign was that Iraq did not use chemical weapons during the war for fear of the Coalition's ability to retaliate with nonconventional weapons.   Moreover, the attacks against Iraq's research, development, and production facilities began the process of taking away Baghdad's ability to threaten its neighbors. Finally, the attrition of artillery among Iraqi frontline units made it very difficult to coordinate a systematic use of chemical munitions. In sum, "even though air attacks against Iraq's chemical-warfare capabilities fell well short of destroying them completely, it by no means follows that these attacks were militarily futile or served no purpose."[10]

---

[9]U.S., Department of the Air Force, *Gulf War Air Power Survey: Summary Report*, pp. 80–81, and *Gulf War Air Power Survey: Operations and Effects and Effectiveness*, Vol II, Part 1, pp. 230–32.

[10]U.S, Department of the Air Force, *Gulf War Air Power Survey: Summary Report*, p. 81.

Iraq's Chemical Warfare Program

The Iraqi chemical and biological warfare program served two specific purposes. Tactically, chemical and biological weapons provided a means to counter numerically superior enemy forces. Strategically, chemical and biological weapons were seen as weapons of last resort for retaliation against a nuclear attack on Baghdad or other actions that threatened the regime. Commanders were pre-delegated the authority to launch long-range (Al Hussein) missiles with biological and chemical warheads in event that Baghdad was hit by nuclear weapons. For the Iraqi leaders, calling biological and chemical weapons the "poor man's atomic bomb" was more than a smart journalistic phrase.[11]

Iraq began research into the production of chemical weapons in the 1970s and by the 1980s began to batch produce selected chemical agents. Although heavily dependent on precursor chemicals from foreign suppliers, Iraq started producing blister agent mustard (HD) in 1981. The quality of the mustard agent was only about eighty percent pure, although it could be either stored for an extensive period in bulk form or weaponized. The nerve gases tabun (GA) and sarin (GB) entered production in 1984; production modifications resolved stabilization problems that marred early efforts. Both the tabun and sarin produced were poor, having a maximum purity of sixty percent. These problems

---

[11]United Nations, Security Council, *Report of the Secretary-General on the Status of the Implementation of the Special Commission's Plan for the Ongoing Monitoring and Verification of Iraq's Compliance With Relevant Parts of Section C of Security Council Resolution 687 (1991)*, U.N. Doc. S/1995/864 (October 11, 1995), p. 11.

prompted Iraq to refocus its nerve agent research, development, and production to sarin (GB/GF). A binary approach was adopted to weaponization to overcome this problem: precursor chemicals for sarin were stored separately for mixing in the munitions just prior to use. This production program resulted in near pure sarin for delivery against Iraq's enemies.[12]

The Al Muthanna State Establishment topped the list of the twenty-one facilities making up Iraq's chemical warfare complex. Also known as the State Enterprise for Pesticide Production (SEPP), the Al Muthanna production complex included the three intended precursor production sites at Al Fallujah and the munitions stores at Al Muhammadiyet. This five-by-five-kilometer facility was Iraq's primary chemical weapons research, development, and production facility, and it is located near the town of Samarra, about seventy-five miles northwest of Baghdad. The site operated continuously from 1983 to 1991, producing thousands of tons of precursors, nerve agents and mustard gas. Chemical agents from this facility included mustard gas, sarin, tabun, and VX.

Al Muthanna was heavily bombed during the Gulf War. From 1992 to 1994, the UNSCOM Chemical Destruction Group operated at this site to eliminate remaining precursor materials, destroy production plants and equipment, and hydrolyze or burn remaining chemical warfare agents.

---

[12]United Nations, Security Council, *Report of the Secretary-General on the Status of the Implementation of the Special Commission's Plan for the Ongoing Monitoring and Verification of Iraq's Compliance With Relevant Parts of Section C of Security Council Resolution 687 (1991)*, U.N. Doc. S/1995/284 (April 10, 1995), pp. 11–12.

Al Fallujah I, located about thirty-seven miles northwest of Baghdad, was under construction when the Gulf War occurred and therefore had not been used for the production of chemical weapons-related items. This facility was intended to be an additional precursor facility for the chemical weapons program.

Al Fallujah II, located about forty miles northwest of Baghdad, produced chemical weapon precursors destined for the Al Muthanna site. Products included chlorine, phosphorus trichloride and oxychloride, thionylchloride, and, with high probability, two direct nerve agent precursors. The site was heavily bombed during the Gulf War. Remaining precursors and equipment were transferred to the Al Muthanna site for destruction under the supervision of the Chemical Destruction Group. UNSCOM established the Chemical Destruction Group in June 1992 and included twenty-three persons, including medical support, from twelve countries.

Al Fallujah III, located some forty-three miles northwest of Baghdad, was in the late stages of construction at the time of the Gulf War. The facility, intended to support the Al Muthanna site, contained multi-purpose production plants. These production areas were extensively damaged in the 1991 bombing. The intended products of this site remain unclear, but it may have been connected with the VX program.

Muhammadiyet, located about eighty-seven miles west of Baghdad, was the primary storage area for filled chemical weapons. At the time of the Gulf War, the site contained numerous chemical warfare munitions, many of them filled with chemical agent. The site was heavily damaged during the war.

The UNSCOM teams and the Chemical Destruction Group completed the elimination of chemical weapons that survived the bombardment.[13]

The inspection results of remaining storage sites, including airfields, are summarized in Table 4–1. In addition, some thirty suspected chemical storage bunkers should be added to the potential target list. The location of Iraq's twenty-one primary chemical weapons sites are shown in Figure 4–1.

The Al Muthanna State Establishment was the primary site used by UNSCOM inspectors for destruction of all chemical activities. With the exception of on-site destruction of some unsafe 122 mm rockets at Kahamisyah, all destruction activities occurred at Al Muthanna. Unfilled munitions and emptied munitions, after being thoroughly decontaminated, were destroyed through crushing or cutting with a oxy-acetylene torch. Filled munitions were either drained as in the case of aerial bombs or destroyed by a combination of simultaneous explosive opening and high-temperature incineration. Bulk mustard agent was destroyed by incineration. The nerve agents GB and GB/GF mixtures were destroyed by controlled hydrolysis in a specially constructed plant meeting UNSCOM specifications. The aqueous

---

[13]United Nations, Security Council, *Report By the Executive Chairman of the Special Commission Established By the Secretary-General Pursuant to Paragraph 9 (b) (i) of Security Council Resolution 687 (1991)*, U.N. Doc. S/23165 (October 25, 1991), pp. 26–27 and U.N. Special Commission, unpublished information paper (October 16, 1995).

Table 4–1
Chemical Facilities

Name	Also Known As	Location	Purpose
Al Muthanna State Establishment	State Enterprise for Pesticide Production (SEPP), Samarra	Near town of Samarra, 75 mi NW of Baghdad	Declared by Iraq as its sole chemical weapons research, development, production and filling facility. Some chemical weapons munitions and bulk agents also stored here. Two types of poison gas manufactured: sarin and mustard gas. Site is 5 km square.
Al Fallujah I	Habbaniyah	37 mi NW of Baghdad	Under construction intended for precursor production. Never completed—had not been used for production of chemical-related items. Extensively damaged in Gulf War.
Al Fallujah II	Habbaniyah	40 mi NW of Baghdad	Special refinery and other CW-related facilities. Produced precursors like phosphorous thichloride, phosphorous chloride, and thionylchloride. Commenced production of significant quantities of chlorine in mid-1990. Extensively damaged in Gulf War.
Al Fallujah III	Habbaniyah	43 mi NW of Baghdad	Used for the formulation of pesticides. Never used for production of chemical agent precursors. Extensively damaged in Gulf War.

Al Fallujah Proving Ground		Storage of 6,394 mustard-filled 155 mm artillery shells.
Al Matasim Aerodome (airfield auxiliary to Al Bakr)		Storage of twenty-five type 250 gauge aerial bombs and 135 type 500 aerial bombs filled with mustard agent. Originally declared by Iraq to be at Al Bakr, but were found at Al Matasim, which is 19 mi north of the base.
Al Bakr Air Base		Aerial bombs stored at auxiliary airfield.
Al Nassariya	An Nasiriyah	One of Iraq's largest chemical weapons dispersal locations.
Al Tuz Airfield		Stored mustard-filled bombs in the open; 122 mm rockets too unsafe to move.
Al Walid Air Base		Stored binary chemical munitions; R-400 bombs.
Basra Petrochemical Complex No. 1		Ethylene production facility (needed to make thiodiglycol and produce mustard gas). Designed to produce 410,000 tons of ethylene products per year. Started in late 1970's.
Chemical Storage Bunkers	Throughout Iraq	Thirty suspected chemical storage bunkers. Eight at Samarra, eight in northern Iraq near Kirkuk and Qayyara.

Continued on next page

Table 4–1 *Continued*

Name	Also Known As	Location	Purpose
Dujayl	Al Dujayl, Al Awarah	3348N/04415E	Storage of thirty chemical-filled ballistic missile warheads. Fourteen of which were of the so-called binary type (a mixture excluding the DF component), the other sixteen were filled with a mixture of nerve agents GF and GB and fifty-six plastic containers filled with DF.
Ibn al Baytar			Contains CW equipment evacuated from Al Muthanna area prior to Gulf War. Pharmaceutical facility, still under construction, intended to produce active ingredients for the Samarra Drug Company.
Jaber bin Haythan	SAAD-24	Adjacent to Badush Dam near Mosul	Produces chemical warfare protective gear.
Khamisiyah	Kamisiyah Military Base	3047N/04626E	CW munitions dispersal area; unsafe 122 mm rockets.
Mosul			Concealed CW production equipment. Mechanical press to make chemical bombs moved to a sugar factory here.
Muhammediyat Stores	Muhammidiyat Stores	87 mi W of Baghdad	Main CW munitions storage facility; unsafe 122 mm rockets.

Qadisiyah Air Base	Al Asad	3347N/04226E	Aerial CW munitions store.
Rutbah	Al Rutbah		Produced acids and other chemical components before Gulf War.
Saddam Air Base			Storage site for mustard-filled bombs; stored in the open.
Salman Pak			Research, production, and storage facilities suspected of being involved with biological weapons. Major CW and BW research center, reached full capacity production in late 1980's (5 tons of mustard and 2.5 tons of sarin per day).
Taji	Al Taji		Originally declared in connection with ballistic missiles, not for chemical weapons. Approximately 6,000 empty aluminum containers intended for filling with nerve agents and insertion into 122 mm rocket warheads were found.
Tammuz Air Base	Al Taqqadum	Habbaniyah	Storage of 200 mustard-filled aerial bombs.

Figure 4–1
Twenty-Three Iraqi Chemical Facilities

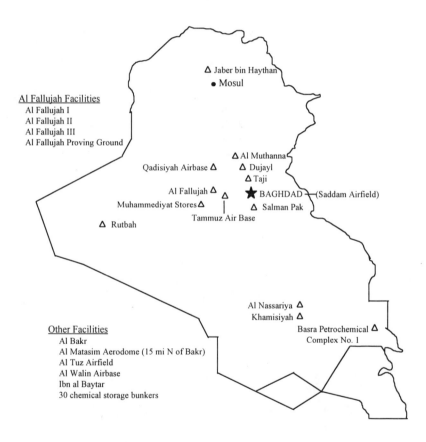

<u>Al Fallujah Facilities</u>
Al Fallujah I
Al Fallujah II
Al Fallujah III
Al Fallujah Proving Ground

△ Jaber bin Haythan
● Mosul

△ Al Muthanna
Qadisiyah Airbase △   △ Dujayl
                         △ Taji
Al Fallujah △   △   ★ BAGHDAD —(Saddam Airfield)
Muhammediyat Stores △ | △ Salman Pak
                    Tammuz Air Base
△ Rutbah

Al Nassariya △
Khamisiyah △

Basra Petrochemical △
Complex No. 1

<u>Other Facilities</u>
Al Bakr
Al Matasim Aerodome (15 mi N of Bakr)
Al Tuz Airfield
Al Walin Airbase
Ibn al Baytar
30 chemical storage bunkers

wastes from the plant were allowed to partially evaporate and the remainder was cast into concrete blocks and buried on the site.[14]

The destruction activities were extensive and lasted for two years. All told, more than 480,000 liters of chemical warfare agents, more than 28,000 chemical munitions and nearly 1,800,000 liters, more than 1,040,000 kilograms and 648 barrels, of some forty-five different precursor chemicals for the production of chemical warfare agents were destroyed.[15] See details in Table 4–2.

Implications for Counterproliferation Operations

The Iraqis apparently began thinking about the need for chemical weapons as early as the 1960s. The Egyptian use of mustard gas and phosgene to terrorize and break the spirit of the Yemeni tribesmen were lessons not lost on Baghdad. A few years later when it came to light that Israel had nuclear weapons during the 1973 war, the Iraqis began to build a strategic rationale for chemical weapons. Having developed chemical weapons by

---

[14]United Nations, Security Council, *Fourth Report of the Executive Chairman of the Special Commission Established By the Secretary-General Pursuant to Paragraph 9 (b) (i) of Security Council Resolution 687 (1991), On the Activities of the Special Commission*, U.N. Doc. S/24984 (December 17, 1992), pp. 22–23.

[15]United Nations, Security Council, *Ninth Report of the Executive Chairman of the Special Commission Established By the Secretary-General Pursuant to Paragraph 9 (b) (i) of Security Council Resolution 687 (1991), On the Activities of the Special Commission*, U.N. Doc. S/1995/494 (June 20, 1995), p. 5.

Table 4–2
Chemical Warfare Items Destroyed at Al Muthanna

Description	Amount
Munitions:	
122 mm warheads	6,773
155mm artillery shells	12,849
Al Hussein warheads (extended-range Scud)	29
Bombs	8,398
Chemical Warfare Agents:	
Mustard agent	398,046 liters
Nerve agent (GA)	21,365
Nerve agent (GB/GF)	61,633
Total Precursor Chemicals	1,798,513 liters
Total Chemical Precursors	1,040,836 kg
Other:	
White phosphorus	648 barrels
Bulk storage containers	32

Source: United Nations, Security Council, *Seventh Report of the Executive Chairman of the Special Commission Established By the Secretary-General Pursuant to Paragraph 9 (b) (i) of Security Council Resolution 687 (1991), On the Activities of the Special Commission*, U.N. Doc. S/1994/750 (June 24, 1994), pp. 25–27.

the early 1980s, Iraq unleashed extensive use of poison gas after it had suffered major defeats in the war against Iran in the spring of 1983. Tactical uses of Iraqi mustard gas and the nerve agent tabun first occurred in 1984. When the Iranians achieved battlefield victories and entered Iran's territory, the Iraqis are said to have readied the deadly nerve gas sarin for use, if necessary.[16]

---

[16]Robert Mandel, "Chemical Warfare: Act of Intimidation or Desperation?," *Armed Forces & Society* (Winter 1993), pp. 201–03.

The evidence now suggests that the Iraqi risk calculus in the Gulf War viewed chemical weapons as those to be used as a last resort or in response to nuclear attacks. It was equally likely at the outset of the Gulf War, and so assumed by the Western Coalition, that Iraq would use its large chemical arsenal tactically against Coalition forces in the field and strategically through attacks against populated areas. This operating premise proved wrong but that does not mean similar inhibitions to chemical weapon use will be present in future crises.

Iraqs' failure to use chemical weapons may have had more to do with the collapse of its command and control system and destruction of much of its delivery capability than with Baghdad being deterred or making the strategy choice to use them only as a last line of defense. If this is the case, as Dr. Andrew Rathmell argues, "Operation Desert Storm may have demonstrated to other proliferators the need to acquire a more sophisticated, flexible and decentralized CW capability."[17]

American policymakers cannot be certain of which strategic course the countries holding chemical weapons will follow. The tactical use of chemical weapons to obtain military advantages on the battlefield seems a likely course of events in many plausible scenarios, especially in those cases in which the enemy cannot strike back with weapons of mass destruction. Similarly, a plan of attack against urban centers to coerce and terrorize civilian populations is another plausible possible course of action as long as there is little or no price to pay in terms of retaliation with nuclear, biological or chemical arms.

---

[17]Dr. Andrew Rathmell, "Chemical Weapons in the Middle East: Lessons From Iraq," *Jane's Intelligence Review* (December 1995), p. 560.

Chapter 5
Developing a Ballistic Missile Production Base

"A Scud missile . . . was responsible for the single worst loss of American forces in the war when it hit an American barracks in Dharan [Saudi Arabia] killing 28 American soldiers and wounding 97 others."[1] The Iraqis launched eighty-eight modified Scud missiles at targets in Israel, Saudi Arabia and Bahrain during the forty-three day, January-February 1991 Persian Gulf War. Some of their warheads were swatted down by Patriot anti-missile missiles, others slammed into residential areas in Tel Aviv and Riyadh, and a few turned and twisted on an uncertain trajectory and finally tumbled harmlessly into uninhabited areas. None of the missiles was armed with radiological, biological or chemical warheads. Except for the warhead that hit the American barracks in Dharan, the Iraqi ballistic missile attacks were militarily pathetic. Politically, however, the attacks were dynamite as the Coalition air forces were compelled to divert a substantial number of intelligence and attack resources from the air campaign to counter-Scud operations. In this regard, the Scud attacks were among the most militarily effective actions taken by the Iraqis.

An essential component of the Iraqi drive to be capable of delivering weapons of mass destruction throughout the Middle East was the effort to build a robust ballistic missile production base. The Iraqis received their first Scud-Bs from the Soviet

---

[1]U.S., Department of the Air Force, *Gulf War Air Power Survey: A Statistical Compendium and Chronology*, p. 542.

Union in the 1970s and later Scud-C derivatives from North Korea.[2] During the 1980s, the Iraqis built a modern missile development and production infrastructure by drawing on foreign technology and expertise, primarily from Brazil, Argentina, Germany, Austria, and Italy. At the time of the Gulf War, the Badr 2000 missile was under development—a 650-mile range, two-staged, solid-fueled missile capable of carrying a warhead in excess of 1,000 pounds. This missile was intended to be Iraq's primary nuclear delivery system and some reports attributed fuel-air explosive warheads with the missile as well. The Badr 2000 was derived from an earlier joint venture with Egypt and Argentina; both partner countries withdrew from the deal with the Iraqis in 1988 and 1989, respectively, under pressure from the United States.[3]

The Iraqis had an ongoing program to extend the range of the Scud-Bs obtained from the Soviet Union. One of these missiles, the "Al Hussein," had been used by the Iraqis during the 1988

---

[2]The Soviet Scud-As were first seen in 1957 and were probably derived in part from the German V-2. Scud-Bs were deployed in the mid-1960s. The missile accompanied Soviet ground forces on an eight-wheeled vehicle called a "transporter erector launcher" or TEL. The Scud-B is a single-stage, liquid-fueled rocket and is believed to have a 300 kilometers (164 miles) range with a payload of 2,100-pounds and an accuracy rate on the order of 3,000 feet. See U.S., Department of the Air Force, *Gulf War Air Power Survey: Operations and Effects and Effectiveness*, Vol. II, Part 2, p. 317.

[3]Michael Eisenstadt, *"The Sword of the Arabs:" Iraq's Strategic Weapon's*, Policy Paper No. 21, (Washington, D.C.: Washington Institute for Near East Policy, 1990), pp. 18–21, and *Like a Phoenix From the Ashes? The Future of Iraqi Military Power*, p. 35. See also Kathleen C. Bailey, *The UN Inspections of Iraq: Lessons for On-Site Verification* (Boulder: Westview Press, 1995), pp. 73–74.

"war of the cities" phase of the Iran-Iraq War. Over a period of eight weeks, the Iraqis launched 189 Al Husseins at six Iranian cities. The Iraqi modifications had doubled the range of the Scud-B with the addition of more than 2,000 pounds of fuel, but at the price of reducing the payload from about 1,800-pounds to 400 pounds. This small, inaccurate warhead offered little capability against military targets unless by chance a direct hit took place. Nonetheless, the warhead was accompanied by the Al Hussein one-ton missile body, which normally slammed into the ground nearby at a terminal velocity of more than four times the speed of sound.

An upgraded variant of the Al Hussein, the "Al Abbas" missile, had a range of about 450 miles and was said to have had greater accuracy than its antecedents. Another variant, one with about the same range as the Al Hussein, was the "Al Hijarah" missile. A few of these were fired by the Iraqis during the Persian Gulf War. In addition to these three ballistic missiles, the Iraqis were working on the "Al Abid" satellite-launch vehicle. The Al Abid consisted of a three-stage missile with a cluster of five Scud boosters strapped together to serve as the first stage, a single Scud rocket engine as the second stage, and a modified surface-to-air missiles (SA-2) as the third stage. Its sole test flight in 1989 was unsuccessful. Finally, the Iraqis were in the process of converting SA-2 missiles to surface-to-surface modes in the "Al Fahd 300" and "Al Fahd 500" programs. (The Bosnian Serbs fired such a weapon against the town of Bihac during a 1994 offensive.) A similar effort was underway to convert the

Silkworm anti-ship missile for ground targets in the "Faw" cruise missile program.[4]

The Iraqi drive to become a regional power also encompassed its "supergun" program or Project Babylon. Drawing on the 1960s joint U.S.-Canadian High Altitude Research Project (HARP) program to test the concept of using gun-assisted rockets to place satellites into earth orbit, the Iraqis were working on a family of superguns capable of firing payloads almost 700 miles away. This was the brainchild of Gerald Bull, a Canadian engineer and ballistics expert. The Iraqis intended to build two guns with a 1,000 mm bore and a 156-meter barrel, and two guns with a 350 mm bore and a 46-meter barrel. Reportedly, the earlier HARP program had demonstrated that the guns could accurately force a rocket-assisted artillery rounds up to 2,000 miles. For the Iraqis, the superguns were cost-effective since they could be built for less than $10 million, and they potentially could sustain rapid refirings. Gerald Bull was murdered in Brussels in March 1990, effectively ending the project. The Iraqis, however, have shown an interest in reviving the supergun program.[5]

---

[4]U.S., Department of the Air Force, *Gulf War Air Power Survey: Operations and Effects and Effectiveness*, Vol. II, Part 2, pp. 317–19; Bailey, *UN Inspections in Iraq*, pp. 73–74; Eisenstadt, *Like a Phoenix from the Ashes*, pp. 35–37; and *Ballistic Missile Proliferation: An Emerging Threat* (Arlington, Va.: System Planning Corporation, 1992), pp. 16–17.

[5]Eisenstadt, *"The Sword of the Arabs,"* pp. 29–31, and *Like A Phoenix From the Ashes?*, pp. 38–39. See also the dramatization of the events leading to the development of the Iraqi supergun program and Dr. Bull's assassination in the movie *Doomsday Gun*.

Like Iraq's biological and chemical program on the eve of the Gulf War, Baghdad's ballistic missile program was well advanced and ready. While nuclear warheads were not yet available, radioactive debris could have been scattered by the missiles. The Iraqis demonstrated extraordinary technical skill and innovation in developing their weapons of mass destruction and missiles. With U.S. and other forces clustered at a few bases, Saudi ports jammed with ships, transport aircraft parked wingtip-to-wingtip, and the Israeli, Saudi, and other Arab peoples open to attack, the region was poised on a precipice of disaster. While the events that unfolded did not trigger the Iraqi use of these weapons, that was a result more of Iraqi strategic doctrine than of a conscious design by U.S. policymakers and military planners.

The Persian Gulf War

Thirty-three of the eighty-eight Scud missile (or their variants) launched during the forty-three days of military operations in the Gulf War occurred during the first week. About half that number were fired during the second week, dropping to five or fewer launches in weeks three and four, and recovering to higher firing rates in the final two weeks.[6] The Scud missile attacks in the first week triggered what was to be called by the American aircrews involved as the "Great Scud Hunt." The Coalition placed 24-hour Scud Combat Air Patrols (airborne attack aircraft) over the known western and eastern Iraq launch areas. The aircraft patrolled these "Scud boxes" and used

---

[6]U.S., Department of the Air Force, *Gulf War Air Power Survey: Summary Report*, pp. 87–88.

surveillance sensors to locate and attack suspected Scud launchers.   Armed reconnaissance aircraft also were used in attempts to locate and destroy the Scuds' mobile launchers and covert pre-surveyed launch sites.[7]

When Saddam Hussein unleashed his Scuds against Israel, he apparently hoped to provoke an Israeli response that would break the alliance of, or cause a serious rift between, the Western and Arab members of the Coalition. If the Israelis had responded, for instance, they would have had to overfly Jordan and Saudi Arabia, potentially triggering significant political fallout. The Bush Administration was determined to avoid taking the bait. Instead, U.S. counter-Scud operations were stepped up as much to keep the Israelis from responding against the Iraqi attacks as they were to find and destroy the missiles and their launchers.[8] The "Great Scud Hunt" would achieve political success in keeping Israel out of the war, thereby preserving the tenuous Western-Arab Coalition. Yet, the price to be paid was in terms of diverting attack aircraft from lucrative priority air strikes to a fruitless chase of Scud launchers. Perhaps this was precisely the effect that Saddam Hussein had hoped for in making his militarily inconsequential Scud attacks.

Prior to the Iraqi attacks on urban areas in Israel, the Scud missiles were viewed by the Gulf War planners as posing little threat to the Coalition. The Scuds were regarded largely as "nuisance weapons" that lacked the accuracy and explosive power

---

[7]U.S., Department of the Air Force, *Gulf War Air Power Survey: A Statistical Compendium and Chronology*, p. 545.

[8]Gordon and Trainor, *The Generals' War*, pp. 230–37.

to do much damage. Hence, planners reduced the number of counter-Scud strikes required by focusing on attacking known locations such as fixed launch sites, support bases, production facilities, and support facilities for mobile launchers. The Scuds themselves were not targeted; such attacks found their way into the "too hard to do" box. In this way, fewer counter-Scud strikes were needed, freeing aircraft to go after more threatening targets in the Coalition strike plan.[9]

Scud missiles, launchers, and production and storage facilities made up the initial target set for ballistic missiles. Intelligence did not have a good handle on the numbers of Iraqi missiles and mobile launchers. Target planners adopted a working estimate of 600 Scud missiles and variants, thirty-six mobile launchers, and twenty-eight fixed launchers in five complexes in western Iraq, plus some training launchers at Al Taji. Since the assumption was that the Iraqis would fire from these fixed locations, the initial Coalition attacks concentrated on eliminating them while plans were being developed for going after the mobile transporter erector launchers (TELs).[10]

If the Iraqis chose to adopt mobile Scud operations, the Coalition air forces were already well schooled in what to expect. Since the Iraqi Scuds initially were supplied by the Soviets and much of Iraq's military training came from the Soviets, it made sense that the familiar Soviet Scud tactics in Central Europe would be emulated in some form by the Iraqis. In this case, it

---

[9]U.S., Department of the Air Force, *Gulf War Air Power Survey: Summary Report*, p. 43.

[10]U.S., Department of Defense, *Conduct of the Persian Gulf War*, p. 97.

meant that several hours of preparation were necessary to launch a missile, which would provide distinct signatures that could be used by the Coalition to locate and destroy the missiles and their TELs. Coalition air planners already trained to deal with the Warsaw Pact in Central Europe understood Scud operations quite well and they anticipated little difficulty in neutralizing them.

When the Gulf War started, the legitimate assumptions on conducting counter-Scud operations were all turned on their heads. Unlike the Soviets, the Iraqis considered the Scud missile (and its variants) to be a strategic weapon rather than a tactical missile organic to ground forces. Strategic operations are planned far more deliberately than tactical operations in the field. First, the Iraqis did not use the fixed locations to launch against Israel or Saudi Arabia; second, they had secret sites prepared for mobile Scud launches that offered few signatures other than the plume of a rocket burning at launch; third, the Iraqis could break down the launch site in ten minutes and scurry away before Coalition fighters arrived; and fourth, they used East German-made Scud decoys to confuse the counter-Scud operations. The mobile TELs were hidden during the day in road culverts, garages and other structures. At night, they would be re-loaded and moved quickly to the prepared covert launch sites, quickly set up and launched against their distant targets. The Iraqis then scurried away before the Coalition's aircraft could target them in the open or on the move.

Coalition air planners faced targeting problems for counter-missile operations similar to those for the nuclear and biological weapons programs. Key components of the target set were not identified prior to the Gulf War. In the case of the ballistic missiles, the missing information included the presurveyed launch sites for Iraq's strategic strikes against Israel, Saudi Arabia, and other Gulf States and the hiding places for the mobile launchers.[11]

Coalition planners and aircrews were baffled by the Iraqi techniques as an ever greater level of effort was made to counter the mobile Scuds. Aircrews reported destroying some eighty Scud launchers but later analysis showed the objects destroyed as having Scud-like signatures and included decoys, tanker trucks, short-range rockets like the FROG-7, and other look-alikes. Special operations forces claimed to have destroyed seven Scuds and called in air strikes on at least five others. Later, it was conceded that most, if not all, of these kills had been decoys. The forces enjoyed a greater success in locating some of the covert launch sites and cutting cables and other actions to disrupt the Scud launches.

Roughly 1,460 Coalition strikes were mounted against the Iraqi ballistic missile target set. Nearly half of these sorties involved dropping ordnance on suspected hiding locations of the mobile launchers and fixed sites. Many highway overpasses, culverts, and other fixed sites took a pounding. Yet, in the end, not a single kill of a mobile Scud launcher could be confirmed. Some 400 strikes were carried out against ballistic missile production and infrastructure. Only 215 strikes were reported to have involved mobile launchers. The 1,000 or so "Scud patrol"

---

[11]U.S., Department of the Air Force, *Gulf War Air Power Survey: Operations and Effects and Effectiveness*, Vol. II, Part 2, p. 330.

sorties dropped weapons on suspected hiding places and other targets besides the Scuds and their mobile launchers.[12]

When all of the evidence was in after the Gulf War, it was apparent that the Iraqis dispersed most of their mobile Scud force in August 1990. The Defense Intelligence Agency discovered many signs of dispersal. Vehicles and activities at the central support bases and other known locations for Scud operations ceased. Yet, the DIA analysts could not find the places to which the Scud vehicles and activities had moved. Hence, carefully couched intelligence reports did not run up a red flag for the target planners and the widespread Scud dispersal went unnoticed. The Iraqis had put into effect excellent concealment and deception measures. Control of the Scud units, for instance, was accomplished exclusively through the use of couriers and landline telephones. The Coalition had little chance of detecting the mobile launchers and Scuds before launch.

The airborne Scud patrols were conducted in the hope that once the launchers were detected the strike aircraft could quickly reach the launch area and pick up the launchers with onboard sensors and destroy them on the scene. The radar and infrared signatures of the mobile launchers could easily be masked by the Iraqis, making them extremely difficult to distinguish from background clutter and other vehicles.

As shown in Table 5–1, the number of Iraqi ballistic missile targets grew over the political time frame.

---

[12]U.S., Department of the Air Force, *Gulf War Air Power Survey: Operations and Effects and Effectiveness*, Vol. II, Part 2, p. 332, and *Summary Report*, pp. 85–86, and Gordon and Trainor, *The Generals' War*, pp. 245–46.

Table 5-1
Known Iraqi Ballistic Missile Targets

Source	Date	Political Time Frame	Targets
U.S. Central Command Air Force Component	June 15, 1990	Pre-war	7
Defense Intelligence Agency	August 2, 1990	Crisis (Operation Desert Shield)	24
Coalition Target Planners	January 16, 1991	War (Desert Storm)	121
U.S. Central Command	July 1992	Post-war	154

Source: U.S., Department of the Air Force, *Gulf War Air Power Survey: Planning and Command and Control*, Vol. I, Part 1 (Washington, D.C.: Government Printing Office, 1993), pp. 189, 193, 199, 201, 212.

These figures tell a story about a lack of focus on Iraq by the intelligence community prior to Saddam Hussein's invasion of Kuwait. Seven targets quickly jumped to twenty-four after the invasion and ended up at 121 at the war's end. Meanwhile, the Coalition's target planners scrambled to assemble one of the most complex air campaigns ever conducted. In the forty-three days of the Gulf War, 41,310 bombing sorties were executed, of which only 1,460 or 3.5 percent were dedicated to counter-Scud operations.[13] The combined effect of an inadequate baseline for targeting to start from, time pressures to develop an intricate air campaign, prudent assumptions that proved faulty, and the Iraqis' concealment and deception in moving to mobile Scud operations, left the Coalition unprepared for the Iraqi missile attacks. The key misunderstanding by the Coalition was in perceiving the Scud as a tactical missile to be used as an adjunct to military

---

[13]Gordon and Trainor, *The Generals' War*, p. 238.

operations. In the eyes of the beholder, the Iraqis, the Scud missiles were strategic weapons to be used to shape the operational and tactical levels of warfare.

Iraq's Ballistic Missile Program

Under the terms of the 1991 United Nations Security Council Resolution 687, the Special Commission (UNSCOM) was given a mandate to obtain an accurate assessment of Iraq's past ballistic missile programs and to destroy or render harmless items prohibited under the U.N. resolution: "all ballistic missiles with a range greater than 150 kilometers, and related major parts, and repair and production facilities." UNSCOM extended major efforts to establish an accurate accounting of Iraq's missile-related activities and dual-purpose capabilities. Obtaining a clear picture of Iraq's missile program was essential to assure that UNSCOM fulfilled its mandate without leaving behind a secret residual missile force.

From the beginning the Iraqis were determined to assure themselves a residual missile force, and presumably biological and chemical, as well as conventional, warheads. Missiles and their components were buried, hidden in a variety of locations, and some reportedly were placed aboard trucks that kept moving around the country. False declarations and blatant lies supplemented the Iraqi actions as a grand cat-and-mouse game was played with the UNSCOM inspectors over the next four years.

In accordance with U.N. Resolution 687, Iraq submitted a declaration in April 1991 that asserted its full disclosure of its ballistic program and related equipment. The first seven UNSCOM missile inspections took place from July 1991 to the end of that year. The Iraqis used procedural delays to disrupt the inspectors, substituted Scud decoys or non-prohibited rockets for the actual Scud missiles, refused to respond to certain requests, made contradictory statements, and generally adopted an uncooperative stance. Nonetheless, the inspectors visited several missile-related sites and ordered the destruction of prohibited items. Bulldozers operated by the Iraqis were used to crush the missile systems and components. Although Iraq had declared that twenty-five of twenty-eight total fixed launch sites in the Western Zone had been destroyed, the United Nations inspectors ordered further destruction. They also discovered several partially constructed fixed launch sites and ordered their destruction.[14]

By the end of 1991, the Special Commission on-site inspections had uncovered no information that clearly contradicted Iraq's disclosures of its missile program. Inspection teams had supervised the destruction of sixty-two ballistic missiles, thirty-three warheads, eighteen fixed missile launch pads, a variety of missile support equipment, and components of Iraq's supergun program (Project Babylon). Yet, important

---

[14]United Nations, Security Council, *Report of the Executive Chairman of the Special Commission Established By the Secretary-General Pursuant to Paragraph 9 (b) (i) of Security Council Resolution 687 (1991)*, U.N. Doc. S/23165 (October 25, 1991), pp. 31–33.

questions remained unresolved over whether Iraq still had ballistic missiles, and its plans for future missile programs.[15]

In February 1992, the U.N. Special Commission ordered the Iraqis to begin destruction of their ballistic missile production program. The Iraqis refused to comply and continued to assert that they no longer had ballistic missiles with a range in excess of 150 kilometers. It took the threat of renewed hostilities with the United States and Britain to get the Iraqis to comply, beginning at the end of March. At the same time, the Special Commission presented incontrovertible evidence that Iraq's initial declarations in April 1991 had not included numerous ballistic missiles and related equipment in its possession. The Iraqis responded by making a new declaration on March 19, 1992, that admitted they had failed to declare ninety-two ballistic missiles and a great deal of associated equipment and vehicles, including mobile missile launchers. They declared that these items had been destroyed unilaterally in the summer of 1991.[16]

The Special Commission dispatched an inspection team to Iraq from March 21–30, 1992, to verify the claims of unilateral destruction of the ninety-two missiles. The missiles had been

---

[15]United Nations, Security Council, *Second Report By the Executive Chairman of the Special Commission Established By the Secretary-General Pursuant to Paragraph 9 (b) (i) of Security Council Resolution 687 (1991)*, U.N. Doc. S/23268 (December 4, 1991), p. 6.

[16]United Nations, Security Council, *Report on the Status of Compliance By Iraq With the Obligations Placed Upon It Under Section C of Security Council Resolution 687 (1991) and Resolutions 707 (1991) and 715 (1991)*, U.N. Doc. S/23993 (May 22, 1992), pp. 5–6. See also Bailey, *UN Inspections in Iraq*, p. 78.

buried at four locations; they were unearthed and their serial numbers taken to match against the list of rocket motors that had been supplied by the Soviet Union. Only eighty-three of the ninety-two missiles could be accounted for. In the meantime, the destruction of Iraq's ballistic missile production facilities and equipment listed in February moved forward.[17]

Nonetheless, the Iraqis continued to interfere with the inspection activities, including tampering with equipment designated for removal for further analysis and doctoring of documentation. The Special Commission conducted inspections to investigate Iraq's ability to produce ballistic missiles, missile guidance-and-control systems, other essential components, and fuels for ballistic missiles.[18]

Discussions with high level Iraqi officials in November 1993 and March 1994 resulted in Iraq's furnishing of additional details on foreign acquisition of critical ballistic missile items as well as its expenditure of ballistic missiles during the war with Iraq and the Persian Gulf War. Iraq submitted documentation covering the

---

[17]United Nations, Security Council, *Third Report By the Executive Chairman of the Special Commission Established By the Secretary-General Pursuant to Paragraph 9 (b) (i) of Security Council Resolution 687 (1991)*, U.N. Doc. S/24108 (June 16, 1992), pp. 23–25.

[18]United Nations, Security Council, *Fourth Report of the Executive Chairman of the Special Commission Established By the Secretary-General Pursuant to Paragraph 9 (b) (i) of Security Council Resolution 687 (1991), On the Activities of the Special Commission,* U.N. Doc. S/24984 (December 17, 1992), pp. 20–21.

period 1977 through December 1990 that accounted for about three-quarters of its prohibited missiles.[19]

The Special Commission conducted fifteen missile inspections in 1994, more than the combined number of inspections undertaken in 1992 and 1993. By the end of the year, the Executive Chairman, Rolf Ekeus, was able to report to the Security Council: "As a result of inspections, lengthy discussions with Iraq's authorities and other bodies and detailed analysis, the commission now possesses a much fuller and accurate picture of Iraq's past prohibited missile programmes than that presented by Iraq in its official 'full, final and comprehensive report' submitted in May 1992."[20] Nonetheless, the cat-and-mouse game with the Iraqis continued into 1996. Iraqi contradictions concerning their missile program continue to appear, and they remain steadfast in their refusal to confirm information obtained by the Special Commission—at least until a preponderance of evidence is placed before them.

In the meantime, UNSCOM set up more than thirty facilities to be monitored by cameras and other sensors to ensure that the Iraqis were not trying to reacquire prohibited missiles. Nearly

---

[19]United Nations, Security Council, *Seventh Report of the Executive Chairman of the Special Commission Established By the Secretary-General Pursuant to Paragraph 9 (b) (i) of Security Council Resolution 687 (1991), On the Activities of the Special Commission,* U.N. Doc. S/1994/750 (June 24, 1994), pp. 12–13.

[20]United Nations, Security Council, *Eighth Report of the Executive Chairman of the Special Commission Established By the Secretary-General Pursuant to Paragraph 9 (b) (i) of Security Council Resolution 687 (1991), On the Activities of the Special Commission,* U.N. Doc. S/1994/1422, Appendix I (December 15, 1994), p. 1.

200 specialized and dual-purpose items were tagged by the Commission. The U.N.'s monitoring teams periodically check Iraq's use of the equipment. More than 1,300 missiles with a range of 50–150 kilometers, non-prohibited missiles, have been tagged and are checked regularly to ensure that they have not been modified for longer range.[21]

By April 1995, the Commission had put a multi-level and comprehensive monitoring system into place covering Iraq's missile research, development, testing and production facilities, as well as those related to dual-use capabilities. The Special Commission missile monitoring group resident in Baghdad provides the surveillance required. The sense at UNSCOM was that it had "essentially completed the accounting of proscribed ballistic missile capabilities," although "investigation will continue until the Commission is satisfied that it has obtained as detailed a picture as possible of all aspects of Iraq's past programmes and current capabilities."[22]

Table 5–2 provides a summary of Iraq's ballistic missile facilities gleaned from a variety of sources. The UNSCOM inspection reports were extraordinarily close-mouthed about the location of the on-site inspections and their findings. The listing depicted in Figure 5–1 at least gives a glimpse into Iraq's

---

[21]*Ibid.*, pp. 3–4.

[22]United Nations, Security Council, *Ninth Report of the Executive Chairman of the Special Commission Established By the Secretary-General Pursuant to Paragraph 9 (b) (i) of Security Council Resolution 687 (1991), On the Activities of the Special Commission,* U.N. Doc. S/1995/494 (June 20, 1995), pp. 4–5.

Table 5–2
Iraq's Ballistic Missile Facilities

Name	Also Known As	Location	Purpose
Al Anbar		50 mi NW of Baghdad	Missile research facility.
Al Dujayl			Storage for chemical-filled ballistic missile warheads.
Al Farouq Factory at Darwah		Outskirts of Baghdad	Manufactured transporter-launchers for the long-range missile program; completely destroyed in the Gulf War.
Al Kinde	SAAD-16; Al Mosul	3 mi N of Mosul	Al Kinde Missile Development Center; conducted research and development, rocket testing, missile test bays, high speed wind tunnels, and climate test chambers.
Al Qa Qaa		24 mi S of Baghdad	Explosive filling of long-range missile warheads.
Al Radwan		Batra	Scud assembly plant.
Al Rafah		37 mi SW of Baghdad	Research, development and testing on rocket motors, static testing for indigenous liquid engines.
An Anbar		Near Karabala	Space research center; launch and testing facility.
As Salman			Intelligence reports of deployed Scud missiles.
Baghdad (Saddam Airfield)			Fixed missile launchers.

Balad Al Suhada	Taj Al Marik, Balat Al Shuhada	30 mi SW of Baghdad	Development of the BADR-2000 missile; no reduction of solid rocket motor fuel, motorcase.
Daura			Fixed missile launchers.
Dhu Al Fiqar	Al Amir	48 mi SW of Baghdad	Motor casing manufacture for BADR-2000 missile; bombed during the Gulf War.
Habbaniyah		Falliyah	Solid Motorcase Production Facility
H-2 Airfield		Western Iraq	Six fixed missile launch positions oriented toward Syria and Israel.
Ishuayb Al Awat			Seven fixed missile launch positions.
Karama		NW outskirts of Baghdad	Research and development on indigenous guidance and control systems.
Karbala	DO-3, Project 1157		Rocket motor test facilities; missile test area.
Luadl Ar Ratqa			Seven fixed missile launch positions.
Mahmudiya	Project 96, Al Hillah	Al Hillah (S of Baghdad)	Rocket propellant production facility, chemical and fuel facility.
Nassr State Establishment		6 mi N of Baghdad	Large machine tool and metal working facility for Scud modification and range extension.
Qasr Amit East			Seven fixed missile launch positions.
Qasr Amit West			Seven fixed missile launch positions.

Continued on next page

Table 5–2 *Continued*

Name	Also Known As	Location	Purpose
Rasheed Camp			Location of missiles, components, and support equipment.
Taji	Baghdad Military Installation, Project 144	18 mi N of Baghdad	Primary location for Iraq's indigenous long-range missile program; air frame design, construction and modification, liquid fuel rocket engine development and production.
Wadi Al Jabaryah			Seven fixed missile launch positions.
Wadi Amil			Seven fixed missile launch positions.
Wadi Hawran			Seven fixed missile launch positions.
Yawm al Azim		39 mi S of Baghdad	Part of Balad Al Suhada group of facilities responsible for development of BADR-2000 missile; plant contained static test equipment and other development technologies.
Zawr Hawran			Seven fixed missile launch positions.

Figure 5–1
Twenty-Nine Iraqi Scud Facilities

extraordinary ballistic missile infrastructure and production base and prepared launch sites.

By mid-1995, it had become apparent that the Iraqis had resumed their acquisition of foreign equipment, technologies, supplies and material for both missile- and non-missile related activities. Iraq explained that these efforts were mostly in support of its development of its Ababil-100 program, a missile with a range of 100 to 150 kilometers.[23] At the same time, satellite pictures showed rebuilding efforts at several key missile facilities, including the Al Kinde Missile Research and Development Facility at Mosul.[24]

The UNSCOM inspections revealed that the Iraqi ballistic missile force was much larger than was believed during the Gulf War. Iraq's evasions throughout more than four years of inspections and supervised destruction of missiles and missile-related equipment raise the question of whether Baghdad had retained a residual force of missiles. Former CIA Director Robert Gates estimated the residual force to number perhaps in the hundreds of missiles. R. James Woolsey, Gates' successor, said that the residual force numbered 100–200 missiles and 12–20 launchers, including Scuds and extended range versions such as

---

[23]United Nations, Security Council, *Report of the Secretary-General On the Status of the Implementation of the Special Commission's Plan for the Ongoing Monitoring and Verification of Iraq's Compliance With Relevant Parts of Section C of Security Council Resolution 687 (1991)*, U.N. Doc. S/1995/864 (October 11, 1995), p. 13.

[24]James Bruce and Barbara Starr, "US Exploits Images of Military Rebirth… As Iraq Rejects UN Resolution on Oil Sales," *Jane's Defence Weekly* (May 6, 1995), pp. 4–5.

the Al Hussein.[25] The U.N. Special Commission in March 1996 estimated that Iraq possessed from six to sixteen ballistic missiles capable of striking Saudi Arabia, Kuwait, and Israel with highly lethal nerve agents and biological weapons. The U.N. inspectors believed the missiles were stored on trucks that shuttled between military installations.[26]

At the same time, Iraq was reported to have twenty-nine FROG-7 transporter-erector-launchers readied at Taji in southern Iraq. The FROG-7 rocket is unguided and solid-fueled; it has a range of about forty-three miles, which is sufficient to strike key targets in Kuwait and Saudi Arabia. A modified version, the "Laith," is claimed to have a longer range.[27]

Implications for Counterproliferation Operations

The Coalition's counter-missile operations during the forty-three days of the Persian Gulf War leave little doubt that intelligence preparation of the battlefield in peacetime is a prerequisite to success during hostilities. Intelligence on key components of the target country's infrastructure and production base for weapons of mass destruction and missiles provides vital input for counter-missile planning, including indications about when and where to attack. The poor record of the Coalition in

---

[25]David C. Isby, "The Residual Iraqi 'Scud' Force," *Jane's Intelligence Review*, 7–3 (March 1995), pp. 115–17.

[26]R. Jeffrey Smith, "Iraq Is Hiding 6 to 16 Scuds, U.N. Suspects," *Washington Post* (March 21, 1996), pp. A1, A28.

[27]Isby, "Residual Iraqi 'Scud' Force," pp. 115–17.

killing Scuds during the Gulf War was due to the inability to detect their whereabouts. Good intelligence preparation could have helped solve the Scud problem.

Iraq absorbed very effectively what its Soviet military advisers taught in earlier years about frustrating an enemy's ability to detect mobile targets. It is reasonable to expect that other countries that maintained close military relationships with the former USSR and now have ballistic missiles would implement the same types of measures that Iraq took to hamper the Scud hunt. These countries include Syria, Libya, North Korea, possibly Algeria, and Iran. Among the practices the Iraqis learned from the Soviets in how to deny locational cuing to missile launchers and applied in the Gulf War are:

- Preparation of ready-hide positions near launch points
- Good concealment methods
- Rapid redeployment
- False targets
- Decoys
- Other camouflage, concealment and deception
- Good operations security, especially radio silence to deny signals intelligence opportunities.

Because of the political sensitivities and potential military ramifications associated with the use of ballistic missiles, as illustrated when Iraqi Scuds struck Israel, the United States must be prepared to act quickly to counter an enemy's decision to launch such missiles. Fusing intelligence and operations together can provide the time-urgent data that are essential for any response to enemy actions. Compressed timelines of ballistic missile attacks, especially those involving weapons of mass destruction, mandate a closer degree of integration of intelligence and operations planning in peacetime than traditionally has existed.

The years beyond 2000 will witness the emergence of new capabilities for the enhancement of peacetime intelligence and operations planning. More accurate and versatile sensors will be developed. Better data fusion capabilities will lead to improved databases on ballistic and cruise missiles and nuclear, biological, and chemical weapons that can be updated rapidly and continuously to assist the intelligence preparation process in peacetime, crisis, and conflict. The marked improvement in the real-time and near-real-time intelligence data will make counter-missile operations more effective.

Counter-missile operations cannot realistically be expected to prevent all missile launches, some of which might involve the use of weapons of mass destruction against friendly populations. Does this anticipated lack of total success mean that counter-missile operations should not be conducted? In the absence of an agreed upon yardstick to assess the value of counter-missile operations, such attacks are open to criticism and debate. Yet, in the final analysis, preventing the launch of even a few missiles armed with nuclear, biological, and chemical weapons could save thousands, even hundreds of thousands, of lives. In this regard, the Air Force survey of its own counter-missile operations during the Gulf War overly focused on the weaknesses and missed the point on the strengths of these operations. The "Great Scud Hunt" was much like rushing the quarterback in football. Our team did not achieve any "sacks" (Scud launcher kills) but, like in a good defense, the "rush" of the airborne Scud patrols most certainly produced some "hurries," "knockdowns," and "batted balls." The disruption may have interrupted Iraqi salvo missile launches, giving the Patriot defensive systems greater chances of success.

The need to better integrate the U.S. intelligence and operations in peacetime for rapid counter-missile operations is one of the foremost lessons that should have been learned from the Persian Gulf War.

Chapter 6
Deceiving Western Export Controls

Iraq's weapons of mass destruction and ballistic missile programs were based on a massive international deception effort designed to mask extensive purchases of foreign commodities and equipment. Through the use of middlemen, fronts, and ownership of foreign companies, the Iraqis were able to avoid detection and ensure uninterrupted supplies. Foreign suppliers competing for larger portions of the Iraqi market were inhibited from sharing information about Baghdad's activities. Elaborate cover stories exploited loopholes in Western export controls by providing plausible civil purposes for purchases of dual-use[1] commodities and technologies. Different components of sensitive equipment were ordered in different countries to get them past customs officials. Components often were designed to look like something other than their true purpose. In the end, there was no one putting it all together. Baghdad had played a masterful hand.

The Iraq case is important not only in pointing the way ahead in the global proliferation of weapons of mass destruction and ballistic missiles, but also in showing how foreign technology assisted Baghdad's clandestine programs. While the Iraqis were self-sufficient in chemical weapons production, imports of essential components provided crucial inputs to their nuclear, biological, and ballistic missile programs. International export

---

[1]"Dual-use" means having both commercial and defense applications, whether a technology, process, product or commodity.

control regimes designed to block the transfer of commodities and technologies essential to developing weapons of mass destruction and their delivery systems slowed and complicated the Iraqi efforts but did not stop them.

What does the Iraqi case tell us about future proliferation and the military challenges that will face the United States? Will the growing globalization of commerce and the spread of knowledge make proliferation easier in the years ahead? Are Western export controls adequate to restrict commodities and technologies that can be applied to weapons programs in Third World countries? What new approaches to U.S. military doctrine and strategy may best respond to the proliferation challenges looming on the horizon of the early 21st century? The answers to these questions may unfold slowly or surprise proliferators may burst upon the international scene to respond more quickly. We are uncertain. But preparing for that uncertainty requires a readiness to adapt our political and military responses to future proliferation as it unfolds.

International Export Controls

The United States and other industrialized countries concerned about the proliferation of weapons of mass destruction have entered multinational agreements to restrict the export of certain commodities and technologies. These export controls, as demonstrated in the Iraqi case, are not foolproof. Yet, they offer the international community's best chance to slow the spread of the requisite technologies for weapons of mass destruction and buy time to encourage a reversal of the drive toward proliferation.

Technologies that provide chokepoints—those steps that are particularly difficult for proliferators to complete—might offer the best pressure points to control proliferation. Those relatively easy steps or technologies that are widely known and used, on the other hand, would make poor candidates to control the spread of WMD and ballistic missiles. It is especially important to possess a good understanding of which dual-use technologies and products have legitimate civil applications as well as those involved in weapons proliferation.[2]

Perhaps the best case example of dual-use sales to Iraq before the August 2, 1990, invasion of Kuwait is provided by the United States itself. The Department of Defense has not made any foreign military sales to Iraq since 1967, and export licenses for commercial sales of defense items have been denied since 1980. Until the Persian Gulf War, the one exception was for items used for the protection of the head of state, Saddam Hussein. On the other hand, U.S. sales to Iraq of dual-use items were not constrained by national security controls, and there were few foreign policy restraints on such trade. All told, 771 licenses for exporting $1.5 billion of dual-use items to Iraq were approved between 1985 and 1990, while thirty-nine applications were rejected. Between 1983 and 1990, the State Department approved nineteen commercial sales to Iraq of items on the U.S. munitions list related to the protection of the head of state.[3]

---

[2]U.S., Congress, Office of Technology Assessment, *Proliferation of Weapons of Mass Destruction,* p. 32.

[3]U.S., General Accounting Office, *U.S. Military Items Exported or Transferred to Iraq in the 1980s,* GAO/NSIAD-94-98 (February 1994), pp. 1-5.

Nuclear Export Control Regimes. The Zangger Committee was formed in the early 1970s to establish guidelines for the export control provision of the Non-Proliferation Treaty (NPT) (Article III (2)). Twenty-nine member states work together in developing a list of controlled items, the "trigger list." These are items normally used in peaceful nuclear programs that, if misused, could contribute to a nuclear explosive program: plutonium, highly enriched uranium, reactors, reprocessing and enrichment plants, and "especially designed or prepared" material equipment for such facilities. Since adopted in 1974, the Zangger Committee has updated the trigger list to provide greater specification in the key areas of the nuclear cycle such as enrichment, reprocessing and heavy water production. To be eligible for export, items on the trigger list must satisfy three criteria: (1) not be used for nuclear explosives, (2) be subject to IAEA safeguard in the recipient non-nuclear state, and (3) not to be reexported unless they are subject to safeguards in the receiving state.

The Nuclear Suppliers Group (NSG), formed after India's 1974 nuclear explosion, was to expand beyond the controls of Zangger Committee and involve the key non-NPT supplier, France. By 1978, the NSG had grown to fifteen countries; the current membership numbers thirty-one, including France. The NSG developed export guidelines that go beyond the Zangger Committee: (1) an agreement between IAEA and the recipient state requiring the application of safeguards on all fissionable materials in its current and peaceful activities, (2) physical protection against unauthorized use of transferred materials and facilities, and (3) restraint in exports that could contribute to the acquisition of plutonium or HEU by a state of proliferation concern. The NSG held annual plenary sessions, the first in more than a decade, to strengthen the regime by upgrading its fuel-cycle control list and through adoption of an arrangement for controlling nuclear-related dual-use items.[4]

Biological and Chemical Export Control Regimes. The twenty-nine member Australia Group is an informal forum of states whose goal is to discourage and impede the proliferation of biological and chemical weapons. The Group harmonizes national export controls and information about countries of concern on chemical weapons precursor chemicals, biological weapons pathogens, and chemical and biological weapons dual-use production equipment. In June 1993, the Australia Group's common export controls were strengthened by the adoption of comprehensive control lists and enforcement measures:

- Biological Agents for Export Control List: viruses, rickettsiae, bacteria, genetically modified micro-organisms, toxins
- Biological Agents Warning List: viruses, bacteria, genetically modified micro-organisms, toxins
- Control List of Plant Pathogens: bacteria, genetically modified micro-organisms
- Control List of Animal Pathogens: viruses, bacteria, genetically modified micro-organisms
- Items for Inclusion in Awareness Raising Guidelines: bacteria, viruses

---

[4]U.S., Arms Control and Disarmament Agency, "Multilateral Nuclear Export Control Regimes," *Fact Sheet* (Washington, D.C.: Office of Public Information, January 24, 1994).

- Control List of Dual-Use Biological Equipment: complete containment facilities, fermenters, centrifugal separators, cross-flow filtration equipment, freeze-drying equipment, independently ventilated protective full or half suits, Class II biological safety cabinets or isolators, aerosol inhalation chambers
- Control List of Dual-Use Chemicals: Commercial and Military Applications.[5]

The difficulties in controlling exports that can be used in biological and chemical weapons programs are increasing as numerous small supplier companies, which are subject to little direct control, including potential suppliers in central Europe, participate in global commerce. The chance of detecting and stopping small amounts of illegal exports is slight since customs authorities are able to check a small proportion of the commodities, equipment, and technologies crossing national borders. These problems are especially severe in biological weapons exports because much microbiological research is conducted by small companies. The illegal export of small amounts of material, such as cultures, will be difficult to prevent. These cultures and related dual-use technologies also could be sent abroad for legitimate peaceful research and development.[6]

These qualifications concerning export controls do not argue for an abandonment of international cooperation on restricting

---

[5]U.S., Arms Control and Disarmament Agency, "Materials Use in the Manufacture of Chemical and Biological Weapons," *Fact Sheet* (October 25, 1993).
[6]Dr. Richard Latter, "The Increased Danger of Biological Weapons Proliferation," *Jane's Intelligence Review* (February 1994), pp. 93-95.

certain exports. Rather, they encourage a realistic perspective and the realization that the Australia Group's control mechanism may be the best we can do. The problem is that they may not be sufficient to block proliferators hell-bent on obtaining biological or chemical weapons.

Ballistic Missile Export Control Regime. The United States has been involved in international efforts to block the spread of missiles since the prospect of their proliferation became manifest. Missiles are fast, difficult to detect, and can pose the threat of surprise attack. Some countries can arm them with biological and chemical warheads and a few with nuclear warheads. Other countries may develop ballistic and cruise missile delivery capabilities.

In response to these developing threats, the United States has pursued discussions with other supplier countries on the importance of controlling the export of missile technology and has served as the driving force behind the formation of the 1987 Missile Technology Control Regime (MTCR). A set of common export guidelines are agreed to by the twenty-five members countries; each of the partner nations implement the guidelines within its own national legislation. The guidelines restrict the transfer of unmanned vehicles capable of delivering a minimum payload of 500 kilograms (1,100 pounds) to a distance of at least 300 kilometers (190 miles), their components and related technology, and, since March 31, 1992, all unmanned vehicles believed to be intended for the delivery of weapons of mass destruction.

The MTCR members countries began implementation of new extended guidelines to have begun in January 1993. The Equipment and Technology Annex is a control list of commodities and technologies divided into two categories:

- Category I - complete rockets and unmanned air vehicle systems and their major subsystems and specially designed production facilities and equipment. A strong presumption of denial applies to transfers of Category I items; transfers of production facilities are banned; and if Category I items are transferred, the recipient must guard against misuse.
- Category II - a number of items make up this class: propulsion components, missile structural composites, flight instruments, inertial navigation equipment, software, production equipment, and related technology.

Despite these export bans and controls, the number of countries capable of producing ballistic missiles indigenously will increase over the next decade. One problem is that space launch vehicles are being developed and tested by several countries. The technology, equipment, and facilities to build space launch capabilities are essentially the same as those needed to produce ballistic missiles.[7]

---

[7]U.S., Executive Office of the President, *Annual Report On the Proliferation of Missiles and Essential Components of Nuclear, Biological, and Chemical Weapons* (March 1993), pp. 8-12, 19-21, 25-27, and U.S., Arms Control and Disarmament Agency, "The Missile Technology Control Regime," *Fact Sheet* (November 7, 1994), pp. 1–2.

Iraq's Foreign Procurement Programs

The multiple foreign sources supplying Iraq's WMD and missile programs are believed to have numbered in the hundreds. Their objective was to obtain the machines and tools necessary to produce their own components. Sometimes the national export controls in place worked. In 1990, for example, American and British customs agents pulled an elaborate sting after an Iraqi-born British businessman sought to purchase from a firm in California forty capacitors that could be used as triggers on nuclear devices. In a later case, German customs officials seized a Swiss shipment of "gear forgings" destined for Baghdad. The parts were critical components for gas centrifuges; the Swiss manufacturer denied any knowledge of their potential military applications.[8]

For every Iraqi foreign procurement stopped, many more got through. Table 6-1 provides a sample of the sources of Iraqi suppliers. One of the most outstanding examples of Iraq's achievements in circumventing Western export controls was its purchase of components for its EMIS program. The Iraqi Atomic Energy Commission developed a large and highly successful procurement network. Other Iraqi state establishments often were used as buyers and contractors in contacts with foreign sources. Orders for equipment (especially manufactured components), were placed directly with foreign manufacturers and indirectly through foreign intermediaries. Multiple pieces of equipment were obtained both directly and indirectly. Indigenous

---

[8]Smith and Frankel, "Saddam's Nuclear-Weapons Dream," p. A45.

Table 6-1

Sample Foreign Sources of Iraqi Supplies

Country	Equipment, Components, Commodities, Technologies and Training
Brazil	Natural uranium ore; design plan for a plant to process uranium; training for Iraqi nuclear personnel
Britain	Milligrams of plutonium; machines for shaping metal into centrifuge parts; industrial vacuum equipment (joint venture with firm in Liechtenstein); resin-mixing and discharge machine to support electromagnetic uranium enrichment; training for Iraqi nuclear personnel
Finland	Copper coils
France	Research nuclear reactors; electron-beam welder to assemble centrifuges; nuclear reactor control panels, instruments, and computers
Germany	Machines to shape metal into centrifuge parts; mass spectrometers to monitor bomb-fuel production; machines for milling metal; jet-molding machine to make centrifuge motors; oxidation furnaces for making centrifuge parts; computers and instruments capable of analyzing metals and powders for atomic bomb manufacture; electronic and computing equipment to measure neutrons and gamma rays; cylindrical presses; training for Iraqi nuclear personnel
Italy	Plutonium separation utility; coordinate-measuring machines to monitor centrifuge production; machines for milling metal; 1.7 tons of enriched uranium

Japan	Carbon fiber; mainframe computer used to process atomic bomb codes
Liechtenstein	Industrial vacuum equipment (joint venture with British firm)
Niger	Natural uranium ore
Poland	Training for Iraqi nuclear personnel
Portugal	Natural uranium ore
Spain	Machines for milling metal
Sweden	Mixer-settler units to extract plutonium
Switzerland	Machines that shape metal into centrifuge parts; electric frequency converters to power atomic bomb fuel production; valves for production of atomic bomb fuel; frequency converters capable of powering centrifuges; training for Iraqi nuclear personnel
United States	Mass spectrometers to monitor bomb-fuel production; valves for production of atomic bomb fuel; computers, electronic testing machines; computer graphics equipment and frequency synthesizers; computers and instruments capable of analyzing metals and powders for atomic bomb manufacture; computers for a nuclear weapons testing site; electronic and computing equipment to measure neutrons and gamma rays; powder press suitable for compacting nuclear fuels; computers to run machine tools
Yugoslavia	Electrical equipment; building materials

Source: R. Jeffrey Smith and Glenn Frankel, "Saddam's Nuclear-Weapons Dream: A Lingering Nightmare," *Washington Post* (October 13, 1991), p. A45, and Diana Edensword and Gary Milhollin, "Iraq's Bomb--An Update," *New York Times* (April 26, 1993), p. A17.

manufacturing capabilities were developed to complete the production of some items.

A large West European foundry, for example, completed the casting and rough machining of six large iron components for the EMIS program; the order was received from the Iraqi State Electric Establishment. At about the same time, the foundry received an order from a European company for twenty-eight large iron pieces. Six of the pieces had identical specifications as those ordered by the Iraqi State Electric Establishment, twelve pieces were halves along a horizontal plane, and ten pieces were various parts of horizontal and vertical components. The pieces were shipped to a German seaport where they were redirected to their final destination. It turns out that these pieces were pre-machined cores for the 1200 mm double-pole magnets destined for the EMIS facility at Tarmiya. The final machining of these components was accomplished at the Aqba bin Nafi State Establishment at Al Radwan. The workers completing the final machining had no idea of what the equipment was for or the identity of the customer. The people that showed up with the components and specifications removed everything when the work had been completed.

A large number of the components for the gas centrifuges also came from foreign sources. Much of the equipment ordered was multi-use in the sense that it could be used for a number of manufacturing processes. Among the most important equipment were a flow forming machine, an electron-beam welder with a special fixture for holding the rotor tube during welding, a large oxidation furnace with special temperature control features, high frequency converters capable of operating a number of large centrifuges, horizontal and vertical balancing machines, quantities of a nuclear grade fluorinated vacuum pump oil, and hand-operated, pneumatic, and electrically controlled bellows valves.[9]

Kathleen C. Bailey of the Lawrence Livermore National Laboratory chronicles how the Iraqis used a variety of ruses to mask their biological, chemical, and missile programs to circumvent international export controls. Her analysis, as summarized in Table 6-2, is particularly useful in showing how well-meaning foreign suppliers entered transactions with Baghdad that were perverted to provide inadvertent assistance to Iraq's secret weapons programs.

The Iraqis have gone to extraordinary measures to hide the identification of their foreign suppliers, including destruction of records and documents, filing serial numbers off of equipment, and refusal to cooperate with the U.N. Special Commission. Identification of foreign suppliers, especially for Iraq's dismantled nuclear program, is important lest Baghdad be able to revive its drive to obtain a full complement of weapons of mass destruction. The Iraqis still have the scientists and engineers and many of the machine tools necessary to restart prohibited programs.[10]

Iraq resumed its effort to acquire foreign equipment in support of its missile program. Using middlemen and front companies,

---

[9]United Nations, Security Council, *Report of the Eighth IAEA On-Site Inspection in Iraq Under Security Council Resolution 687 (1991)* (November 11-18, 1991), U.N. Doc. S/23283 (December 12, 1991), pp. 10-13.

[10]U.S., Foreign Broadcast Information Service, JPRS Report, *Proliferation Issues* (February 5, 1993), pp. 21-22 and (March 22, 1993), pp. 26-27.

Table 6-2
Circumvention of Biological, Chemical, and Missile Export Controls

Weapons	Circumvention Measures
Biological	Interest in medical and biotechnology research provides cover for biological weapons program; France designed factory in Daura for manufacturing vaccine for foot-and-mouth disease; numerous German companies "supplied microbiological parts and fluid culture media;" assorted serums, heating equipment, drying closets, fermenters, incubators, and other items suitable for biological weapons were exported to Iraq; the U.S. Public Health Service, Centers for Disease Control, and companies sent strains of tularemia bacterium and various toxins and bacteria for the purpose of medical research.
Chemical	Iraq depicted its purchases of chemical precursors and equipment as intended for industrial applications in the pharmaceutical, fertilizer, pollution control, pesticide, and plastics industries; end use for chemical-related exports are impossible to verify because of multi-uses; great difficulties are confronted in trying to trace chemical precursors, especially when Iraq had commodities shipped to and from one, two and even three countries before reaching their destination; German companies built Iraqi pesticide and other plants that were diverted to chemical weapons production; companies from the United States and other members of the Australia Group supplied chemical weapons-related products and equipment.
Ballistic Missiles	Iraqi ballistic missile development programs began in the 1980s before the inception of the MTCR; many Iraqi suppliers were not members of the MTCR; Iraq sent its missile designers and engineers abroad for training; national bureaucracies often did not enforce the MTCR guidelines; front companies were used to facilitate foreign procurements, often through re-export arrangements; illegal actions by Western businessmen circumvented national laws and the MTCR; Iraqi offices managing foreign buys often were located in the countries of non-MTCR members; substitute materials were sometimes used.

Source: Kathleen C. Bailey, *The UN Inspections in Iraq: Lessons for On-Site Verification* (Boulder: Westview Press, 1995), pp. 96-106.

Iraq resurrected its covert network of purchasing agents to buy missile parts from Russia and firms in Europe. Among the equipment obtained by Iraq are accelerometers and gyroscopes, specialty metals, special machine tools, and a technologically advanced furnace that can be used to manufacture missile engine parts. These activities are underway despite the fact that sales of industrial equipment and technology, especially those intended for military use, are strictly prohibited by the U.N. Security Council resolutions following the Gulf War. Americans are among those conspiring with the Iraqis. The U.S. Customs Service had some thirty-five cases open in October 1995 investigating violations of the Iraqi trade sanctions. Connie Fenchel, head of U.S. Customs Strategic Investigations Office, says that: "The Iraqis are in fact continuing, if not enhancing, their efforts to procure missile technology and every other type of technology, in fact anything that's currently embargoed."[11]

Implications for Counterproliferation Operations

The New Era following the Cold War is witnessing an explosion of global trade on a massive scale. Advanced information technologies are spreading worldwide, creating new relationships between suppliers and recipients. Previous U.S. controls on telecommunication technologies have been slashed. At the same time, the export control system designed to contain

---

[11]As quoted in R. Jeffrey Smith, "Iraq Buying Missile Parts Covertly," *Washington Post* (October 14, 1995), pp. A1, A20. See also Thalif Deen, "UNSCOM Accuses Iraq of Purchasing Arms," *Jane's Defence Weekly* (October 21, 1995), p. 5, and James Bruce, "Playing Hide and Seek With Saddam," *Jane's Defence Weekly* (January 3, 1996), p. 15.

Soviet military power has been dismantled. The United States has proposed a global multilateral regime to control arms and dual-use technologies. Two major goals of the prospective regime will be to prevent destabilizing buildups of weapons by establishing a formal process of transparency, consultation and restraints in the export of weapons and other sensitive dual-use technologies to designated states (Iran, Iraq, Libya, and North Korea). The "New Forum," as it is called, will require responsible national transfer policies, adherence to non-proliferation norms, and a commitment to effective export controls.[12]

Will the export controls be sufficient to keep pace with rapidly increasing foreign trade and at the same time prevent proliferation of weapons of mass destruction and missiles? The question must remain open since the spread of technology and scientific knowledge continues to open new vistas for countries worldwide. Export controls must remain dynamic to meet changing demands of world trade. In October 1995, for instance, the United States eliminated controls on the export of all computers in North America, most of Europe, and parts of Asia. Computer export controls for other countries will be eased. President Clinton explains that "for the former Soviet Union, China, and a number of other countries, we will focus our controls on computers intended for military end uses or users, while easing them on the export of computers to civilian customers."[13]

---

[12]U.S., Department of State, *Dispatch*, 6-42 (October 16, 1995), pp. 752-54.
[13]U.S., Department of State, *Dispatch*, 6-43 (October 23, 1995), p. 776.

Despite gaping dual-use loopholes such as this latest one with computers, the export control mechanisms will no doubt rack up some impressive successes over the coming years. Yet, they will not, and cannot be expected to have, a perfect record. Proliferation of nuclear, biological, and chemical weapons and cruise and ballistic missiles will remain a most nettlesome problem. The New Era is marked by a great uncertainty.

Chapter 7
The Implications of Regional WMD Proliferation for U.S.
Military Strategy: Assessment

Thus far, we have examined Iraq's drive to field weapons of mass destruction, the U.S.-led effort to eliminate the emerging Iraqi WMD threat during Desert Storm, and the post-Gulf War revelations about the true extent of Iraq's WMD development. This chapter places what we have learned into a proper perspective by answering two main questions: First, how does what we have learned sharpen our understanding of the urgency and magnitude of the threat posed by WMD-armed regional adversaries?[1] And second, what implications does this have for the U.S. military posture?

Lessons Learned From the Iraqi Experience

This case study, unsettling as it may be, provides keen insights for future planning. Using Iraq as a surrogate, U.S. decisionmakers and military planners can now make legitimate assumptions about the regional WMD threat and can begin laying plans. Broadly speaking, four main conclusions can be drawn from the analysis: regional WMD proliferation is no longer

---

[1]Most analyses of the national security and military implications of WMD proliferation focus on either nuclear weapons or chemical and biological weapons, but not both. We chose to consider their implications collectively, primarily because all three allow their owner to tap into the same U.S. vulnerabilities (discussed below), and therefore have similar implications for U.S. military posture.

tomorrow's problem; regional WMD proliferation is likely to continue unabated; monitoring WMD proliferation is extraordinarily difficult, particularly during crises and conflict; and as a distinct target set, WMD programs present a daunting military challenge.

Regional WMD Proliferation Is No Longer Tomorrow's Problem. Before Desert Storm, U.S. policymakers generally believed that regional WMD proliferation could be forestalled. Official U.S. policy toward Iraq reflected this view. Howard Teischer, National Security Council staff member during the Reagan Administration (1982–87), recalls the period:

> A combination of wishful thinking and narrow legalistic analysis had persuaded most policymakers that allegations that Iraq was acquiring weapons of mass destruction were little more than anti-Arab propaganda intended to block U.S.-Iraqi reconciliation. Iraq's signature on various non-proliferation treaties bolstered this belief.[2]

Many national security experts and observers of the world scene considered regional WMD proliferation inevitable, but few expected the problem to rear its ugly head in the near-term. Even as the potential threats to U.S. national security were being reexamined in the wake of the Soviet Union's disintegration, little serious thought was being devoted to the military implications of WMD-armed regional adversaries. The stunning series of post-Gulf War revelations about the size, scope, sophistication and

---

[2]Howard Teischer, "The Naive Hope That Allowed Hussein to Weigh Mass Murder," *Los Angeles Times* (September 10, 1995).

maturity of Iraq's nuclear, chemical and biological weapons and ballistic missile programs—captured in this report—have changed all that. The term "counterproliferation" has moved right along side the less-ominous "nonproliferation" in the official U.S. national security vernacular. This alone is testimony to Washington's acknowledgment that regional WMD proliferation is upon us, and that U.S. strategy must now be reoriented to deal with it.

Regional WMD Proliferation Is Likely To Continue Unabated. Three observations drive this conclusion. First, regional states have powerful incentives to acquire WMD. The Cold War's end has lifted the restraining influence of superpower competition and unleashed long-standing regional tensions that remain unchecked. Rogue states in search of regional hegemony seek WMD to intimidate and/or defeat their neighbors and to deter and, if necessary, disrupt and block U.S. intervention. Indeed, having witnessed the U.S.-led Coalition's systematic destruction of the powerful Iraqi military in 1991, potential U.S. adversaries are now scrambling to acquire a strategic "equalizer." As General K. Sandurji, former chief of staff of the Indian Army, said after the Gulf War, "the lesson of Desert Storm is don't mess with the United States without nuclear weapons."[3]   Fear, however, is not the sole motivator. While Operation Desert Storm unveiled America's powerful military capabilities, it also exposed two serious U.S. vulnerabilities to WMD: one is America's highly-publicized aversion to casualties and the other is its near-absolute dependence upon theater access for projecting decisive

---

[3]An interjection by General Sandurji at the Defense Nuclear Agency's *International Conference on Controlling Arms* (Richmond, Va.: June 7–10, 1993); from the author's notes and later reaffirmed by General Sandurji.

military power. To be sure, as one analyst has noted, a regional aggressor armed with nuclear weapons "may not need to fight the Americans."[4]

Meanwhile, as more and more states proliferate, even non-aggressor states will feel compelled to follow suit to deter WMD-armed aggression. "I think we're going to see the number of nuclear weapons states grow exponentially over the next 10 to 25 years," said Dr. John Hassard, a British nuclear physicist who teaches nuclear nonproliferation courses at the University of London. Hassard said that, as one country goes nuclear, "its five or six neighbours will feel much greater pressure to acquire nuclear weapons."[5] Nowhere is this domino-like trend more apparent than in Southwest Asia, where, according to Lewis Dunn, a former assistant director at the Arms Control and Disarmament Agency:

> [The] nuclear status quo . . . is becoming increasingly unstable. . . . Until quite recently, it was widely assumed that Israel's nuclear monopoly would persist for the foreseeable future. Radical countries such as Libya, Syria, or Iraq were presumed to lack the technical skills and organizational capacity to acquire nuclear weapons, and in any case Israeli military action was seen to offer a coercive nonproliferation *deus ex machina*. These assumptions now need to be reassessed in light of

---

[4]Patrick Garrity, "Implications of the Persian Gulf War for Regional Powers," in *Weapons Proliferation in the 1990s,* ed. by Brad Roberts (Cambridge, Mass: MIT Press, 1995), p. 45.

[5]"Nuclear Watchdog Unleashes New Powers to Stall Evaders," *Jane's Defence Weekly* (April 8, 1995), p. 23.

revelations about Iraq's nuclear program, the Gulf War bombing of Iraq, and new incentives for proliferation among countries within the Middle East.[6]

The Iraqi case also shows that proliferation is all but impossible to prevent. Perhaps the most disturbing lesson we have learned is just how little effect Western nonproliferation efforts really have on a determined proliferator. Iraq, a signatory of the 1968 Nuclear Non-Proliferation Treaty, obtained much of the technology, components and materials necessary to develop WMD by manipulating the many dual-use loopholes in Western export controls. Export controls may help to slow proliferation, but a country dedicated to obtaining WMD has many tools at hand to circumvent Western customs barriers. Many suspected proliferators remain outside established WMD control regimes, and not all signatories are reliable adherents.[7] And, while our experience with Iraq may provide a useful sketch of future regional WMD threats, that same information can also aid aspiring proliferators who now know more about how to develop and conceal a WMD program (and protect it from attack).[8]

---

[6]Lewis Dunn, "New Nuclear Threats to U.S. Security," in *New Nuclear Nations: Consequences for U.S. Policy*, ed. by Robert D. Blackwill and Albert Carnesale (New York: Council on Foreign Relations, 1993), pp. 21, 22.

[7]For example, suspected biological proliferators Egypt, Israel, and Syria are non-signatories of the Biological Weapons Convention (BWC). Meanwhile, convention signatories China, Cuba, India, Iran, Iraq, Laos, Libya, Myanmar, North Korea, Russia, South Africa, and Taiwan are all suspected of having an active biological weapons program. See Holly Porteus, "Grappling With the BW Genie," *International Defense Review* (March 1995), pp. 32-34.

[8]For an interesting look at the Iraqi deception, see David A. Kay, "Denial and Deception Practices of WMD Proliferators: Iraq and Beyond," in *Weapons Proliferation in the 1990s*, ed. by Brad Roberts, pp. 303–25.

Finally, nuclear, chemical and biological weapons and missiles (ballistic and cruise) are getting easier to develop and acquire. In today's world, the flow of information simply cannot be controlled and the transfer and dispersal of WMD technology is inevitable.[9] Indeed, many of yesterday's "customers"—the early proliferators—are now suppliers. Perhaps most striking about Iraq's progress in developing nuclear weapons is that it all occurred prior to the collapse of the Soviet Union and was accomplished with largely antiquated and improvised technology. The Soviet collapse introduces a new wrinkle into the proliferation problem. Today, many rightfully fear that nuclear expertise, advanced nuclear technology, high-grade nuclear materials, and possibly even full-up nuclear weapons will be available to the highest bidder from the cash-strapped former Soviet republics, possibly through organized crime syndicates peddling WMD on the international black market.[10]

Monitoring Regional WMD Proliferation Will Be Extraordinarily Difficult, Particularly During Crises And Conflict. One of the first lessons the Coalition partners learned after the smoke cleared in the Gulf was that their pre-war WMD-related intelligence had been woefully inadequate. The Coalition air campaign planners, for example, targeted just eight Iraqi

---

[9]For a good discussion of this phenomenon, see Michael Moodie, "Beyond Proliferation: The Challenge of Technology Diffusion," *Washington Quarterly* (Spring 1995), pp. 183–205.

[10]Phil Williams and Paul Woessner, "The Real Threat of Nuclear Smuggling," *Scientific American* (January 1996), pp. 40–44; Theresa Hitchens, "U.S. Lawmakers Want Tighter Lid on Nuke Material," *Defense News* (March 18–24, 1996), p. 10; and Tim Weiner, "U.S. Vulnerable to Terrorist Chemical Weapons," *New York Times* (March 21, 1996), p. A5.

nuclear facilities during the war; in contrast, International Atomic Energy Agency inspectors discovered some fifty-six such sites afterward. Before Desert Storm, experts pegged Iraq as several years from fielding even a crude nuclear device; it is now believed they were just twelve to eighteen *months* away from producing one or more nuclear devices and/or deliverable weapons.

Obviously, these facts suggest that far more resources should now be devoted to WMD-related intelligence gathering and that substantial innovation in intelligence collection and analysis is required. It would be wrong, however, to cast full blame on the intelligence community. Iraq went to extraordinary lengths to shroud the development of its WMD triad. The highly-compartmentalized Iraqi nuclear program was designed from inception to shield progress from both IAEA inspections and American reconnaissance satellites. Iraq spread critical facilities throughout the country, resulting in fragmented Western intelligence that only made sense in retrospect after intrusive IAEA and U.N. on-site inspections. Moreover, in exploiting technologies considered obsolete by Western standards (such as calutrons used in the uranium enrichment process), Iraq avoided exhibiting many of what were believed to be the "tell-tale" signs of a nuclear proliferator.

As Desert Storm approached, the Iraqis concluded that much of their WMD was vulnerable to air attack and began taking countermeasures to further confound Coalition intelligence. Facilities justifiably considered "fixed" by air campaign planners were effectively relocated by the Iraqis, resulting in attacks against empty buildings after critical components had been moved (and often buried).   The more easily-dispersible assets, such as chemical and biological weapons stocks, and mobile assets, such as Scud missile transporter-erector-launchers (TELs), proved nearly impossible to locate.

The bottom line here is simple, if sobering: there will always be uncertainty when it comes to monitoring potential proliferators, and it will always intensify when the proliferator feels threatened.

As a Distinct Target Set, WMD Programs Present a Daunting Military Challenge. Using the Iraqi WMD program (January 1991) as a baseline target set for future counterforce operations, one is immediately struck by the sheer scale of the projected effort. Planners would have to target over 240 "fixed" facilities (e.g., research, development and production sites), not to mention hundreds of other dispersed (e.g., weapons stocks) and/or mobile assets (e.g., missile launchers). The large size of many of these facilities means that each would have several critical aimpoints. Literally thousands of separate aimpoints, each requiring one or more precision weapons, would need to be attacked—far more than were attacked in the first twenty-four hours of Desert Storm.

Such an operation would be highly complex and very risky, since the proliferator will undoubtedly know America viewed his WMD development as a burgeoning threat. Offensive action would likely come after a prolonged period of rising political tensions, during which the suspect WMD program will likely have received sustained U.S. and international attention. Fearing military action, the proliferator would likely step up defensive preparations, such as concealing and dispersing critical WMD assets and placing air defenses in a much higher state of readiness.[11] Many of the most critical facilities could very well be located in hardened and/or deeply-buried bunkers resistant to all but the most advanced penetrating weapons and virtually invulnerable to current-generation cruise missiles. Many facilities would probably be guarded by an overlapping system of sophisticated local-area and terminal air defenses. Moreover, many could be located in, or indeed relocated to, heavily-populated urban areas. "Hugging" civilians with WMD plays against the well-known Western aversion to collateral damage, especially when played back by CNN to the court of world opinion.

If one assumes that the objective of any U.S. counterforce operation would be to neutralize a proliferator's WMD capabilities, the Gulf War air campaign sets a rather weak precedent. By the Air Force's own admission, the forty three day assault—conducted with impunity and under optimal conditions—merely "inconvenienced" the Iraqi nuclear effort.[12] Meanwhile, Iraq's chemical and biological programs had fully matured prior to Desert Storm, and large quantities of weaponized biological and chemical assets survived the allied onslaught. Indeed, the real setbacks to the Iraqi WMD program occurred only *after the fact*, when U.N. and IAEA inspectors uncovered the

---

[11]See Philip Zelikow, "Offensive Military Options," in *New Nuclear Nations: Consequences for U.S. Policy*, ed. by Robert D. Blackwill and Albert Carnesale, p. 170.

[12]U.S., Department of the Air Force, *Gulf War Air Power Survey: Operations and Effects and Effectiveness*, Vol. II, Part 2, p. 329.

full extent of Iraq's WMD programs (albeit with many details hidden forever by the Iraqis).

In short, we can expect future WMD target sets to be large, extremely difficult to find, hardened, well-protected, and located next to things or people we do not want to damage or injure. Counter-WMD operations therefore could likely be large in scale, extraordinarily challenging from both operational and logistical standpoints, politically controversial, and very risky. Prudence demands that our strategy for countering the regional WMD threat not rest on our ability to eliminate it militarily.

Implications of WMD-Armed Adversaries on the U.S. Military Posture

Potential U.S. regional adversaries seek WMD not only to intimidate and/or defeat their neighbors, but to deter and, if necessary, disrupt and block outside intervention. Most, if not all, such states view the United States as their principal extra-regional threat, and WMD as the means for "devaluing" U.S. military might by exploiting America's aversion to casualties and its clear dependence upon access to ports, airfields and military facilities in the theater of conflict.

Weapons of mass destruction may enable regional states to limit American military capabilities in four main ways: deterring, or critically delaying, a U.S. decision to intervene; intimidating America's regional allies; disrupting and blocking in-progress U.S. power projection operations; and limiting U.S. war aims.

Deterring U.S. Intervention. Post-Cold War experiences in the Persian Gulf, Bosnia, Somalia and elsewhere have plainly illustrated the American public's concern over military casualties. This has fueled the widely-held perception in the developing world that Americans do not have the will to engage in bloody conflicts in which U.S. survival is not at stake. Therefore, an aggressor would seek to deter the U.S. intervention by convincing the U.S. leadership that the expected costs (i.e., casualties) outweigh the possible gains. Ideally, this objective would be met prior to initial U.S. deployments.

Threatening the use of WMD against U.S. assets enroute to, or already in, the theater of conflict is one obvious strategy. Ports, airfields, bases, American troops, and naval forces in the region would present attractive targets. Moreover, an aggressor does not need to actually use, or even threaten to use, his weapons to achieve this objective. As long as the United States is reasonably certain the enemy does in fact possess deliverable[13] WMD, the specter of casualties is likely to dominate the political

---

[13]"Deliverable" WMD is defined here as that which can be used against forward-deployed U.S. forces or U.S. interests and allies abroad. The various delivery platforms include, but are not limited to: the "air-breathers" (WMD-capable strike aircraft and cruise missiles), ballistic missiles, and "non-traditional" delivery means (such as delivery via merchant ships, civilian airliners, and commercial trucks). A more detailed discussion of adversary WMD delivery systems occurs below in the "theater defense" section of Chapter 8. Note: The primary focus here is on WMD in a regional context. Although WMD threats to the continental United States could also affect U.S. decisionmaking in a regional crisis—and possibly inhibiting U.S. power projection capabilities—they are beyond the immediate scope of this analysis.

decisionmaking process.[14] Public debate over potential casualties could very well cause the president to balk at sending American forces abroad or to backpedal after a U.S. deployment has already begun.

A 1993 RAND study on post-Cold War proliferation illustrated the chilling deterrent potential of regional nuclear weapons: "In a war of nerves with a new nuclear adversary, the sense was that the threat posed by a small number of nuclear-armed missiles might negate a U.S. threat to deploy five to ten divisions [to a region]. . . ."[15] The study placed several foreign and security policy experts in simulated war-games based on nuclear confrontations involving North Korea, Iran, India, and Pakistan.   One study participant, Raymond Garthoff of the Brookings Institution, recently recalled the paralyzing effect the threat of nuclear weapons had on decisionmaking:

> Because the risk is clearly so much greater when you confront a nuclear-armed opponent, it tends to make decisionmaking far more difficult. We found that the level of risk inherent when nuclear weapons were involved can push you to take a much weaker, or conversely a far more drastic, course of action than you normally would.[16]

---

[14]See, for example, Edward N. Luttwak, "Toward Post-Heroic Warfare," *Foreign Affairs* (May/June 1995), pp. 109–22.

[15]Marc Dean Millot, Roger Molander, and Peter Wilson, *"The Day After . . ." Study: Nuclear Proliferation in the Post-Cold War World*, Vol. 1, Summary Report, MR-266-AF (Santa Monica: RAND, 1993).

[16]As quoted in James Kitfield, "Counterproliferation," *Air Force Magazine* (October 1995), p. 59.

Intimidating America's Regional Allies. In the post-Cold War era, U.S. military intervention in a regional crisis could hinge on the formation of an international coalition with allies in the theater. Moreover, the success of America's current regional warfighting strategy (deploying large numbers of forces into the region) depends upon access to bases and adjacent waters in the theater. A regional aggressor might threaten WMD use to prevent the formation of such a coalition, splinter an existing coalition, or coerce neighbors into denying the United States access to their ports, airfields, military facilities, and airspace.

Lack of coalition support would increase domestic opposition to the use of American force abroad and could delay a U.S. military response. Should America find itself compelled to intervene unilaterally, power projection operations would still depend upon theater access. Absent the ability to rapidly pour tactical aircraft and ground forces into the theater, America's military power projection capabilities would be seriously circumscribed, especially in the critical, early phases of a conflict. In such instances, the United States would be forced to rely heavily on long-range bombers and sea-based firepower.

Disrupting and Blocking U.S. Deployments and/or Combat Operations. A WMD-armed aggressor could also employ his weapons to cripple on-going U.S. power projection operations. For example, the aggressor could launch chemical or biological strikes against such targets as the main ports and airfields of debarkation to disrupt the flow of U.S. combat aircraft, troops, heavy military hardware, munitions, and other supplies into theater.[17] In his 1995 report to Congress, John M. Collins, a senior defense specialist at the Congressional Research Service, argues that regional aggressors could apply chemical and/or biological weapons to "great advantage." "Results would be devastating," says Collins, "if they immobilized activities at seaports, international airfields and supply depots so the United States could neither introduce forces into the theater rapidly nor adequately sustain U.S. and allied forces already in place."[18]

If launched at the front of a short-warning invasion, such a WMD attack could buy an aggressor additional time to achieve his initial military objectives (e.g., occupation of the victim state's capital city or critical economic assets), dig in and disperse, and force the U.S. to decide between conceding or fighting a much more costly war of eviction.[19] Moreover, the greater the success

---

[17]Germany's December 1943 air attack on Bari Harbor, a strategically vital Allied port in Southern Italy, sets an important—if little-known—precedent in this regard. One of the seventeen ships sunk in the raid was a U.S. merchant vessel carrying 100 tons of 100-lb mustard gas bombs. Many of the bomb casings ruptured, severely contaminating the harbor and causing over 2,000 military and civilian casualties. Bari was closed for twenty-one days and not restored to pre-raid operational capacity for over two months. The resulting supply shortages helped cripple the Allies' winter campaign, allowing the Axis hold northern Italy through the spring of 1944. See: Glen B. Infield, *Disaster at Bari* (New York: Macmillan, 1971).

[18]John M. Collins, *Weapons of Mass Destruction: The Impact of Proliferation on U.S. Military Posture*, Congressional Research Service (Washington, D.C.: Library of Congress, June 2, 1995), p. 27.

[19]Adversary forces on the offensive are typically more vulnerable to attack than when in defensive positions. On the offensive, for example, armored forces, which are essential for seizing territory and vital objectives, are massed and moving (thus providing a high profile signature vulnerable to air attack). Once objectives have been achieved, armored forces can disperse and dig in

an adversary achieves with his initial offensive, the more likely he will have gained something with which to bargain. An adversary who managed to seize the Saudi oil fields or a key city such as Seoul would be in a better position to negotiate with the United States (by threatening destruction of the "hostage") compared to an adversary whose forces had been stopped short of his objectives.

Nuclear strikes by a regional aggressor would, for obvious reasons, be even more devastating. As one Department of Defense official reportedly put it: "In the Gulf War we sent 96 percent of our tonnage through two ports. If Saddam had had three nuclear weapons, he could have destroyed our war-fighting capability."[20]

Limiting U.S. War Aims. Adversaries could also use or threaten the use of WMD to protect their regimes or states from total defeat. In *Nuclear Deterrence in a Regional Context*, RAND analysts Dean Wilkening and Kenneth Watman argue that, in such situations:

> Regional leaders essentially become 'nondeterrable.' Either way, the adversary faces the prospect of losing his hold on power—the ultimate value for many Third World regimes, especially totalitarian regimes. . . . Launching a nuclear attack under these circumstances may appear suicidal to other states, but from the perspective of a

---

to minimize their vulnerability to U.S. counterstrikes (and increase the level of effort required by U.S. and allied forces).

[20]As quoted in Collins, *Impact of Proliferation on U.S. Military Posture*, p. 27.

leader near defeat, threatening to attack may be a risk worth taking if it represents the last hope for survival. . . Hence the opponent's threat to use nuclear weapons first is credible.[21]

Obviously, a regional leader's reliance on WMD in such cases would be intended to raise the perceived costs of defeating him beyond the expected gains. This logic applies evenly to nuclear, chemical and biological weapons—precisely because all are capable of causing "mass destruction."[22] Finally, as Wilkening and Watman point out, it is very important to note that it is the adversary, not the United States, who decides what is, and what is not, regime threatening. This means an adversary has the option of announcing that U.S. action "of any sort" is regime threatening at any point during the crisis.[23] The effects could be chilling to U.S. decisionmakers and the general public.

Conclusions: A Snapshot of the Future

This case study of Iraq's WMD programs provides U.S. policymakers a template for development of the deterrence policies, military strategies, and force structures necessary to deal effectively with the proliferation challenges that lie ahead. The WMD arsenals of tomorrow's adversaries will probably look

---

[21]Dean Wilkening and Kenneth Watman, *Nuclear Deterrence in a Regional Context* (Santa Monica: RAND, 1994), p. 36.

[22]One would, however, expect an adversary in possession of all three to resort first to non-nuclear WMD in the hopes that he could meet his objective without having to test the sensitivity of America's nuclear trigger.

[23]Wilkening and Watman, *Nuclear Deterrence in a Regional Context*, p. 32.

something like Iraq's during the Persian Gulf War. The engineering and industrial process for developing nuclear, biological and chemical weapons, and ballistic missiles, are largely the same in countries worldwide. By using the Iraqi arsenal as a surrogate, one can make legitimate generalizations about the wide-ranging WMD threats and postulate alternative options for dealing with them successfully.

Before moving on, it is perhaps wise to define "success" in this context. Generally speaking, weapons of mass destruction threaten U.S. national security to the extent that they enable regional aggressors to limit America's ability to protect its interests and allies abroad with military power. Success, therefore, in countering the regional WMD threat is defined here as the United States having a military posture (encompassing strategy, doctrine and forces) upon which enemy WMD can have the least possible effect.

Chapter 8
The Implications of Regional WMD Proliferation for U.S.
Military Strategy: Options

In Chapter Seven, we examined the many ways in which WMD-armed regional aggressors might hamstring U.S. power projection capabilities. This final chapter candidly explores U.S. options for dealing with regional WMD proliferation by answering two critical questions: Is the Clinton Administration's program for countering the regional WMD threat up to the task? And, secondly, what alternative policy, strategy, and force structure options are available to U.S. decisionmakers and military planners?

Clinton Administration Policy: The Counterproliferation Initiative

The Clinton Administration recognized early on the seriousness of the regional WMD threat and placed it on the national security agenda. In the Bottom-Up Review (BUR), a comprehensive review of the nation's defense strategy and force structure initiated in March 1993, newly-appointed Secretary of Defense Les Aspin outlined the Administration's initial view of the problem:

Dangers posed by nuclear weapons and other weapons of mass destruction—that is, biological and chemical weapons—are growing. . . . In most areas where U.S. forces could potentially be engaged on a large scale, such as Korea or the Persian Gulf, our likely adversaries already possess chemical and biological weapons. Moreover, many of these same states (e.g., North Korea, Iraq, and Iran) appear to be embarked upon determined efforts to acquire nuclear weapons. Weapons of mass destruction in the hands of a hostile power not only threaten U.S. lives but also challenge our ability to use force to protect our interests.[1]

Secretary Aspin went on to state that while nonproliferation efforts—"primarily diplomatic measures" to "prevent the spread of weapons of mass destruction to additional countries"—would continue, "DoD must also focus on *counterproliferation* efforts to deter, prevent, or defend against the use of WMD if our nonproliferation efforts fail."[2]   Specifically, Secretary Aspin called on the Defense Department to:

- Improve WMD-related intelligence, "both overall [peacetime] WMD threat assessments and timely intelligence and detection to support battlefield operations and management."

- Improve the ability U.S. forces to "seize, disable, or destroy arsenals of nuclear, biological and chemical weapons and their delivery systems."

---

[1]Les Aspin, Secretary of Defense, *Report on the Bottom-Up Review* (Washington, D.C.: Department of Defense, October 1993), p. 5.

[2]*Ibid.*, p. 6 (emphasis in original).

- Maintain "flexible and robust nuclear and conventional forces to deter WMD attacks through the credible threat of devastating retaliation."

- Develop "ballistic and cruise missile [active] defenses," with a goal of *deploying* "advanced theater missile defenses to protect forward-deployed U.S. forces" and *providing the capability* for the "limited defense of the United States."

- Improve passive defenses, including "better individual protective gear and better antidotes and vaccines for our forces in the event they are exposed to chemical or biological attacks."

- Develop "other improved equipment, capabilities, and tactics to minimize the vulnerability of U.S. forces to WMD attacks."

- Develop "better technologies to detect weapons transported covertly into the United States and elsewhere for terrorist purposes."[3]

Less than two months later, Secretary Aspin unveiled the Defense Department's Counterproliferation Initiative in an address to the National Academy of Sciences.[4] Immediately,

---

[3] *Ibid.*, p. 7.

[4] Les Aspin, Secretary of Defense, "The Defense Department's New Nuclear Counterproliferation Initiative" (Address to the National Academy of Sciences, Washington, D.C.: December 7, 1993), as reported in Harald Müller and Mitchell Reiss, "Counterproliferation: Putting New Wine in Old Bottles,"

concern set in—both at home and abroad—over Aspin's apparent emphasis on counterforce. The fear was that America "intended to establish itself as global [WMD] judge, jury, and executioner" and was fielding the means to "unilaterally and preemptively" destroy WMD programs in the developing world.[5]

Foreign anxiety over America's apparent shift toward preemption was multifaceted. Some worried that preemptive U.S. attacks could easily provoke a wider regional war. A perfect example of such sentiments arose in 1993–94 on the Korean Peninsula, where North Korea's neighbors asked the U.S. to refrain from actions that could precipitate armed conflict. There was also concern that such strikes could have devastating collateral consequences for the civilian populations in both the target country and neighboring states. A counterforce strike against an operational nuclear reactor or reprocessing plant, for example, could release massive amounts of radioactive material. Many questioned the international legitimacy of preemptive strikes and expressed alarm over America's apparent creation of an imperialistic "counterproliferation doctrine" that seemed to ignore long-standing international nonproliferation verification and enforcement procedures.[6]

Washington Quarterly (Spring 1995), p. 143.

[5]*Ibid.*

[6]"By verification procedures," we are referring to those activities carried out by international agencies (e.g., IAEA) tasked with verifying compliance to the various WMD non-proliferation treaties (e.g., the Nuclear Non-Proliferation Treaty, Chemical Weapons Convention, and Biological Weapons and Toxins Convention); the U.N. Security Council is the primary international non-proliferation enforcement mechanism.

Developing countries—particularly those in the Arab and greater Islamic world—were particularly concerned about what the U.S. criteria would be for distinguishing "friendly" from "unfriendly" proliferators. Finally, there was concern, voiced primarily by America's NATO allies, that the United States viewed preemption as a substitute for traditional nonproliferation policy. Some alliance members, in fact, sought guarantees from the U.S. that nonproliferation would remain NATO's focus, and that all military action would be in accordance with international law, whereby only a U.N. mandate or a clear case of self-defense could serve as the basis for the use of force.

Over the past two years (1994–96), the Clinton Administration has worked very hard to assure the international community that counterproliferation and preemption are not one and the same, and that the Counterproliferation Initiative involves only the preparation of U.S. forces to deter and, if necessary, defeat an adversary's WMD use on the battlefield.[7] Indeed, in his discussion of the Counterproliferation Initiative in his March 1996 *Annual Report*, Secretary Aspin's successor William J. Perry makes no reference to preventive or preemptive war.[8]

Despite the purported progress, uncertainty still characterizes the Clinton Administration counterproliferation program. Does the Administration view regional WMD proliferation as a truly fundamental change in the security environment, requiring a

---

[7]Müller and Reiss, "Counterproliferation: Putting New Wine in Old Bottles," p. 143.

[8]William J. Perry, Secretary of Defense, *Annual Report to the President and the Congress* (Washington, D.C.: Government Printing Office, March 1996), pp. 53–57.

reevaluation of the way America fights its wars, or as simply a new wrinkle in the old threat, whereby a few minor adjustments will keep the current approach viable? The Defense Department, it seems, views the counterproliferation program as something of a novelty and certainly not as an emerging pillar of the mainstream strategy-making and force-structuring processes. In Secretary Perry's 1996 *Annual Report*, for example, counterproliferation does not show up as a "Defense Component" on a par with the major defense categories.[9] Rather, "Counterproliferation and Treaty Activities" can be found in the *Pursuing Defense Initiatives* section—alongside "Quality of Life," "Economic Security," and "National Security Space Organization and Management." As a result, Congressional support for the counterproliferation initiative has been decidedly lukewarm. Members and their staffs, while cognizant of the need to protect U.S. forces against WMD-armed adversaries, skeptically view the Clinton Administration initiative as a "Christmas tree" upon which the armed services are struggling to "hang" both the current military strategy and the force structure that strategy justifies.[10] Despite claims to the contrary, progress on implementing the Counterproliferation Initiative has been dismally slow. Underfunded and haphazardly implemented, the Counterproliferation Initiative has produced more smoke than fire.

---

[9]Defense Components in Secretary Perry's 1996 *Annual Report* include strategic nuclear forces, land forces, maritime forces, aviation forces, mobility forces, special operations forces, space forces, ballistic missile defenses, C^4I, and the national guard and reserve.

[10]Reiss and Müller, "Counterproliferation: Putting New Wine in Old Bottles," p. 145.

Assessing U.S. Options for Countering WMD-Armed Regional Adversaries

According to Mitchel Wallerstein, deputy assistant secretary of defense for Counterproliferation Policy, the Department of Defense counterproliferation program currently spans two broad "disciplines:" defenses (both active and passive) and counterforce. Furthermore, because the counterforce capabilities for preemptive operations are the same as those required for wartime counterforce operations, it is quite clear, as some Defense officials privately admit, that the Administration's program seeks to provide the implicit option for launching preventive or preemptive strikes against rogue proliferators—official U.S. policy notwithstanding.[11] Finally, the Administration remains firmly committed to the maintenance of a strong nuclear force. Our assessment of military options currently available or under development will therefore include evaluation of the following: nuclear deterrence, theater defenses, and counterforce operations.

Nuclear Deterrence. During the Cold War, America's national security strategy focused on deterring Soviet aggression. Nuclear deterrence—threatening devastating nuclear retaliation to deter a Soviet attack—was relied upon because it was accepted that both sides viewed direct conflict as unacceptably dangerous. The "unacceptable danger" was rapid escalation to a nuclear exchange which both superpowers thought would destroy their respective societies.

---

[11]*Ibid.*, p. 143.

With the Cold War over, U.S. national security strategy is now focused on deterring and, if necessary, defeating regional aggression against its interests and allies abroad. The United States expects its regional adversaries to be armed not only with nuclear weapons, but chemical and biological weapons as well. Can we simply transplant nuclear deterrence into our post-Cold War security strategy? As illustrated in this case study of Iraq's WMD and its implications for U.S. policy, there are many ways in which WMD-armed aggressors can limit America's ability to protect its interests with military power. Here we ask a very simple question: "Can America's nuclear superiority over its regional adversaries mitigate these potential problems?"

Continuing reliance upon nuclear deterrence appeals to those in the military who believe it offers an exquisitely simple solution to a difficult problem. Doing so requires no modification to the current military strategy and has limited implications for current force modernization plans. Nuclear deterrence is also supported by some fairly compelling strategic logic: America's nuclear superiority will provide the U.S. with perpetual "escalation dominance" over potential regional adversaries. That is, the U.S. will always be capable of using nuclear weapons to retaliate to WMD attacks with a level of violence potential adversaries cannot match.

Unfortunately, this may not make much difference to an aggressor facing total defeat at the hands of the United States. How deterrable, for example, would the Soviets have remained in the midst of a preemptive U.S. nuclear strike? Not very. As we learned in Chapter Seven, the U.S. can have no certainty heading into a conflict that an aggressor will not resort to WMD at some point. The aggressor, after all, defines total defeat. Even with limited war aims, the United States could find itself facing a non-deterrable adversary. John R. Powers and Joseph E. Muckerman offer a realistic perspective: "there is no choice but to assume that a political decision to acquire nuclear weapons is also a political decision to use them."[12]

Nuclear deterrence may be even more limited against chemical and biological weapons. Because such weapons do not have to cause large numbers of U.S. casualties to be effective, an aggressor contemplating a chemical or biological attack has good reason to believe the U.S. interests at stake are not as great as his own. Prior to an invasion of a neighboring country, for example, an aggressor could launch chemical or biological attacks against theater facilities critical to the success of U.S. power projection operations. Even if the strikes proved highly advantageous to the aggressor by denying U.S. forces use of the attacked port, airfield or other facility, it is difficult to imagine the United States retaliating with nuclear weapons unless the strikes killed a large number of Americans.

Perhaps most importantly, the concept of deterrence is inherently flawed as a policy for dealing with weapons that the aggressor need not actually employ to achieve warfighting objectives. Nuclear, chemical and biological weapons do not have to be used, or even explicitly threatened, to limit America's ability to project decisive power. The lack of intrinsic U.S. interests in some regions may make the American nuclear

---

[12]John R. Powers and Joseph E. Muckerman, "Rethink the Nuclear Threat," *Orbis* (Winter 1994), p. 103.

deterrent non-credible.   Deterrence can not be guaranteed. American decisionmakers can never be freed from the paralyzing effects of potential WMD use by a regional adversary. If not, how can deterrence be considered a reliable means for countering the regional WMD threat? Keith Payne of the National Institute for Public Policy has an answer: "Prepare for deterrence failure."[13]   A greater dependence on theater defense and counterforce may therefore be indicated for future planning.

Many believe that an implicit American nuclear threat deterred Saddam Hussein from using WMD during the Gulf War, but the post-Cold War utility of nuclear deterrence was not fully tested. Would Hussein have kept his chemical and biological weapons holstered if Coalition forces had marched on Baghdad, thus threatening his very existence? Would Hussein have utilized his chemical and biological arsenals differently if he had expected the U.S. to intervene?[14] If he had anticipated a U.S. response, would Saddam Hussein have deferred the seizure of Kuwait until he had deliverable nuclear weapons? "With nuclear-armed . . . delivery systems it is uncertain whether Iraq would have followed the tradition of nonuse," Germany's Uwe Nerlich observes, "and the lesson from the Gulf War for future nuclear-armed aggressors could produce a fundamentally different strategic behavior."[15]

---

[13]Keith Payne, remarks presented at the National Defense University Foundation and the American Defense Preparedness Association breakfast seminar (Washington, D.C.: Capitol Hill Club, April 10, 1996).

[14]It seems clear that Saddam Hussein was under the impression—albeit a false one—that the United States would not militarily oppose Iraq's "annexation" of Kuwait.

[15]Uwe Nerlich, *Toward a New Nuclear Weapons Regime*, prepared for Sandia National Laboratories (Ebenhausen, Germany: Stiftung Wissenshaft und

The Gulf War raised many more questions about nuclear deterrence than it answered, questions which must be answered before deterrence can be accepted as viable. As the late Les Aspin put it, Cold War deterrence has "no single analogue" in the new era.[16]

Theater Defense. Defense is the second element of the U.S. counterproliferation program. Active defenses are used to prevent WMD from reaching and detonating at their targets; passive defenses are intended to minimize the damage sustained in WMD attacks that cannot be foiled. In crises involving WMD-armed adversaries, such defenses would be used to protect U.S. forces in the field and reassure regional allies whose cooperation the U.S. depends upon. It is generally accepted that neither active nor passive defenses can eliminate the risk of WMD attack. The goal, therefore, is to develop defenses which minimize those risks such that a WMD-armed adversary cannot easily intimidate its regional neighbors or the United States.[17]

Active Defenses. Adversary WMD delivery systems can be broken up into two main categories: "air breathing" vehicles (e.g., WMD-capable strike aircraft and cruise missiles) and

---

Politik, Forschungsinstitut für Internationale Politik und Sicherheit, August 1993), p. 9.

[16]As quoted in Michele A. Flournoy, "Implications for U.S. Military Strategy," in *New Nuclear Nations: Consequences for U.S. Policy,* ed. by Robert D. Blackwill and Albert Carnesale, p. 141.

[17]Wilkening and Watman, *Nuclear Deterrence in a Regional Context*, pp. 46–48.

ballistic missiles.[18] Theater air defense systems—comprising land-based and airborne early-warning radar, fighter interceptors, anti-aircraft artillery, and surface-to-air missile systems—are used to counter the former; theater missile defenses (TMD) are used against the latter.

Air Defenses. Theater air defenses have been protecting forward-deployed U.S. forces several decades now and figure to remain a key element of future U.S. power projection operations. And, although the United States seems reasonably well-prepared to defend against most air-breathing threats, current generation U.S. theater air defenses have yet to be seriously tested.[19] This is important from the American perspective since minimizing the risks of an adversary penetrating air defenses are now much higher: the advent of WMD leaves no margin for error.

---

[18] While the focus here is on traditional delivery systems, "non-traditional" delivery (e.g., via merchant ships, civilian airliners, and commercial trucks) is often presented as the third main category of WMD delivery system. A range of options have been proposed for defense against non-traditional WMD delivery, including, but not limited to enhanced rear-area security, optimization of intelligence capabilities for discerning possible delivery modes and the timing of attacks, and the development of methods for detecting the presence of and rendering harmless WMD aboard otherwise non-threatening vehicles. See Collins, *Impact of Proliferation on U.S. Military Posture*, p. 27; Wilkening and Watman, *Nuclear Deterrence in a Regional Context*, p. 46; and Albert Carnesale, "Defenses Against New Nuclear Threats," in *New Nuclear Nations: Consequences for U.S. Policy*, ed. by Robert D. Blackwill and Albert Carnesale, pp. 204–05.

[19] Iraq essentially conceded air superiority to the allies in the opening hours of Desert Storm and made virtually no attempt to penetrate allied air defenses. See U.S., Department of the Air Force, *Gulf War Air Power Survey: Summary Report*, p. 57.

Historically, the effectiveness of America's theater air defenses has been assessed in terms of the attrition rate imposed on enemy aircraft. Air defenses have not been expected to destroy every enemy plane. Against WMD-capable platforms, however, the stakes are much higher. Even a single breach of the defenses can have devastating consequences. The primary challenge for U.S. defenses will be discerning which of the adversary's aircraft are WMD-capable.[20]

In the case of new nuclear proliferators, the number of nuclear warheads is likely to be small. Moreover, U.S. intelligence will likely be intensely focused on which aircraft are nuclear-capable. Thus, the chances for successful defense against aircraft-delivered nuclear weapons appear to be fairly strong at present. As regional nuclear inventories grow and nuclear weapons get smaller, however, the number of nuclear-capable aircraft may increase and the likelihood of successful interception will inevitably decline, if incrementally.[21]

As we have learned from Iraq, chemical and biological proliferators can quickly develop hundreds of weapons suitable for delivery by many types of strike aircraft. This increases the difficulty of both distinguishing which aircraft are chemical- and/or biological-capable and of shooting down all such aircraft

---

[20]Carnesale, "Defenses Against New Nuclear Threats," pp. 207–08.

[21]As a rule of thumb, nuclear weapon size will decrease as the proliferator becomes more technologically advanced. The smaller the weapon, the greater the number of aircraft capable of carrying them. Basic logic suggests that it will be more difficult to intercept all aircraft-delivered nuclear weapons successfully as the numbers of both warheads and nuclear-capable aircraft increase.

in a multiple-aircraft strike. Accordingly, one can assume that U.S. theater air defenses may be somewhat less effective against aircraft-delivered chemical and biological weapons than they are against nuclear weapons.

For the same reasons cruise missiles are valuable to the U.S. in attacking enemy WMD,[22] they present a more daunting air defense challenge than do strike aircraft. Cruise missiles are relatively small and fly very low, making it hard to distinguish them on radar from other objects at their altitude. Moreover, because they are unmanned, an adversary would likely be more willing to use them (as opposed to manned strike aircraft) against the robust American defenses. Fortunately, cruise missile payloads are limited, so new nuclear proliferators are not likely to soon develop a nuclear warhead sufficiently small to be cruise missile-borne. Chemical and biological warheads, however, can easily be sized for cruise missiles. The U.S. should therefore expect potential adversaries to exploit rapidly-proliferating cruise missile technology to acquire a survivable means of delivering WMD.

Theater Missile Defenses. Theater ballistic missiles represent perhaps the greatest challenge in defending against regional WMD. This is due not only to the wide-scale proliferation of missiles and missile technology, but also to the technical obstacles to developing highly-effective missile defense systems. Accordingly, TMD is receiving perhaps the most attention from the Pentagon of any counterproliferation-related program.[23]

---

[22]Discussed below on pages 189 and 191-92.

[23]Kitfield, "Counterproliferation," p. 58.

Former CIA director R. James Woolsey reminds us that China's 1996 missile tests with impact points near Taiwan serve as a repetition of the lessons learned from Iraq's 1991 attacks against Israel and Saudi Arabia: "Ballistic missiles are being used by hostile states for blackmail, terror and to drive wedges between us and our friends."[24] The warhead accuracies possible using the Global Positioning System, he says, may make it feasible within a few years for Iraqi or Chineses leaders to threaten to destroy, for example, the Knesset or nuclear power plants in Taiwan.[25]

Immediately following the Gulf War, there was a widespread recognition of the need for active defenses to protect against theater missiles by destroying them inflight. These capabilities were to satisfy three main objectives: prevent saturation of the forward combatant commander's "point defenses" or those assigned to protect specific U.S. military assets, neutralize potential warhead effects, and ensure minimal leakage in defense of critical assets. A defense-in-depth concept would provide multiple opportunities to neutralize attacking theater missiles in boost, post-boost, mid-course, and high-altitude interception as well as close-in point defense. This approach would offer multiple opportunities to intercept attacking missiles and/or warheads with differing technology. From the outset, planners

---

[24]R. James Woolsey, "Missile Blackmail Exists Today," *Wall Street Journal* (April 8, 1996), p. A18.

[25]*Ibid.*

recognized that active defenses could consist of space-, air-, ground-, and sea-based systems.[26]

The medium- and long-term TMD forecast remains uncertain, largely because of a shift in Defense Department budgetary priorities. America's TMD *technology* is expected to improve markedly over the medium- and long-term. Multi-tiered systems in development are expected to cover a larger area and be capable of intercepting missiles at longer ranges than point defenses, and reasonably high single-shot kill probabilities are expected (e.g., between 0.6 and 0.8).[27] Much farther off are boost-phase laser interceptors.

None of these systems, however, have been seriously tested, and all are being caught in the budget crunch.[28] A recently completed Defense Department review of the Ballistic Missile Defense Organization (BMDO) led to the deployment slippage of all three core TMD programs: Patriot Advanced Capability (PAC-3) point defense system, Theater High-Altitude Area Defense (THAAD), and new Navy Lower Tier point defense. Despite strong Congressional support, the Navy Upper Tier System was not included on the Defense Department's core list.[29]

---

[26]Ted Gold and Jasper Welch, Co-Chairmen, *Principal Observations and Context*, Theater Defense Architecture Study, Vol. I (Washington, D.C.: Department of Defense, January 1993).

[27]Wilkening and Watman, *Nuclear Deterrence in a Regional Context*, pp. 46–47.

[28]Collins, *Impact of Proliferation on U.S. Military Posture*, p. 24.

[29]See U.S., Congress, House of Representatives, Joint Hearing of the House Military Research and Development and Military Procurement Subcommittees (March 7, 1996).

The deployment of PAC-3 has been slipped from FY 1998 to FY 1999; Navy Lower Tier has been slipped from FY 1999 to FY 2002. And THAAD, until only recently considered by the Department of Defense as the centerpiece of its TMD strategy, received a $2 billion cut over the Future Year's Defense Plan and was slipped from 2000 to 2006.[30] According to Defense Secretary William Perry, the $2 billion reduction in the THAAD program "will be made available for other Department priorities."[31] Meanwhile, the threat is getting more complex all the time; potential U.S. adversaries are expected to develop TMD countermeasures such as decoys and early-release chemical and biological submunitions.[32]

The budget-driven reductions in the U.S. missile defense program opened gaping holes in theater defenses, and it remains highly questionable whether the situation will improve any time soon. Having sacrificed theater missile defenses on the altar of "other Department priorities," the Pentagon disconnected a critical element of a larger defense-in-depth strategy for countering WMD proliferators. The upper tier THAAD system would have provided multiple shots in a wide area defense (100-500 kilometers) against theater ballistic missiles with WMD and submunitions. Boost phase intercept capabilities were also sacrificed. Secretary Perry says, on the one hand, "boost phase defense is *critical* to defending against NBC-armed ballistic missiles," and on the other hand when explaining the budget-

---

[30]*Ibid.* Under the current plan, a fly-off between Navy Upper Tier and THAAD is scheduled for 2004, with only the winner to be deployed.

[31]Perry, *Annual Report to the President and the Congress*, p. 222.

[32]Wilkening and Watman, *Nuclear Deterrence in a Regional Context*, p. 47.

driven restructuring of TMD programs, "the Department will *continue to explore* concepts for boost-phase theater missile defense. . . ." [emphasis added][33] In short, the Clinton Administration's theater missile defense program shows all of the strategic insight of a bookkeeper.

Passive Defenses. Passive defenses are the last resort against WMD attacks that cannot be deterred, preempted or foiled; they are required to minimize the effects of damage caused by air and missile attacks by reducing the enemy's ability to target U.S. forces, lessening the vulnerability of U.S. forces and its supporting infrastructure, and improving the potential for force reconstitution. The passive measures to accomplish these tasks might include efforts to induce enemy targeting errors through one or more of the following: mobility, dispersal, counter-surveillance, electronic warfare, redundancy, deception, and camouflage and concealment. Nonetheless, the increased numbers and lethality—and improved accuracy and effectiveness—of WMD could allow adversaries the capability to overcome these passive measures.[34] Unfortunately, even if such defenses performed beyond current expectations, the enemy could still achieve many of his WMD objectives. Therefore, passive defenses can most usefully be assessed in terms of damage control. There are five main ways to limit damage from a WMD attack: force dispersal, detection and identification, individual and collective protection, decontamination, and casualty management. Current U.S. capabilities are deficient each area.

---

[33]Perry, *Annual Report to the President and the Congress*, pp. 56, 222.

[34]Dudley Tademy, *Passive Defense & ABO Threat,* Theater Defense Architecture Study, Annex G (Washington, D.C.: Department of Defense, January 1993).

Force Dispersal. The most lucrative U.S. targets for WMD attack are likely to be those where large concentrations of U.S. troops, supplies and/or equipment are located. Dispersal of these assets would considerably limit the damage that a nuclear, chemical or biological strike could inflict. Unfortunately, such a multiplication of installations seem to be a prerequisite for any large-scale U.S. power projection operation. One would have to assume that any serious attempt at dispersal would greatly, if temporarily, reduce the military effectiveness of the forces involved.

Detection and Identification. In the wake of WMD attacks, near real-time battlefield detection systems will be needed to warn forces "downwind" of oncoming nuclear fallout and chemical and biological diffusion. During the Cold War, the Department of Defense placed a heavy emphasis on radiation detection capabilities for both land and sea, and they remain largely intact. The same cannot be said for chemical and biological detection systems. Currently available chemical detection devices are effective only at fixed sites; deployed systems cannot consistently and accurately discern agent types, gauge persistence, or measure intensities. Battlefield biological detection systems are non-existent.[35] The Clinton Administration has placed its number one passive defense priority on developing a real-time detection and characterization of biological and chemical warfare agents, including a standoff capability.[36] Yet, many of the promising R&D programs remain underfunded.

---

[35]Collins, *Impact of Proliferation on U.S. Military Posture*, p. 30.

[36]John M. Deutch, Deputy Secretary of Defense, *Report on Nonproliferation and Counterproliferation Activities and Programs* (Washington, D.C.: Government Printing Office, May 1994), p. 31.

Individual and Collective Protection. In the wake of the Gulf War, U.S. chemical and biological defense equipment stocks were rated insufficient, and they remain so.[37] Military forces unable to evacuate prior to a WMD attack today will seek to shield themselves and their supplies and equipment from weapon effects. Blast and fall-out shelters would be critical during nuclear attacks, as will defenses against the accompanying electromagnetic pulse (EMP). During chemical and biological attacks, U.S. forces would need masks, protective clothing, prophylaxis, antidotes, and therapeutic drugs. Simplified tactical protective shelters also are needed. Existing collective protection assets provide limited rest and relief capability, they are not easily deployed, and they offer limited protection against "novel agents."

Decontamination. Current systems are manpower intensive, are based on water technology of washing down equipment, and present a sizable logistics burden. A range of procedural, operational, and technological changes are needed. Inadequate funding of the fixes needed promise a continued use of water-based technology, together with its heavy manpower and logistics burdens, for some time.

Casualty Management. A vastly improved handling system is needed for chemical and biological warfare casualties. For example, casualty vital signs monitors, resuscitation devices, and skin protectants can all contribute to improving casualty management. Funding constraints have limited progress in this

---

[37]Collins, "Impact of Proliferation on U.S. Military Posture," p. 29.

area as well as for other passive defense measures.[38]

Counterforce Operations. Counterforce—using military force in an attempt to disable or destroy an adversary's WMD capabilities before they can be used—is the third "pillar" of the Clinton Administration's counterproliferation program. Officially, the United States is preparing only for wartime counterforce operations. Yet, with WMD proliferation generating greater concern in the armed services and on Capitol Hill, interest in the development of a credible preemptive counterforce capability is growing in some circles.[39] Counterforce operations would commence either before the outbreak of hostilities or in the war's opening moments; waiting any longer would unnecessarily expose one's deployed forces to enemy WMD and defeat the purpose of the strategy.

Theoretically, such a counterforce strategy is sound; neutralized WMD cannot be used against U.S. forces or its regional interests and allies. But is a counterforce strategy realistic? Can U.S. decisionmakers be confident enough of America's wartime capability to "defang" a WMD-armed adversary so as to act decisively during crises? The U.S.-led Coalition's 1991 counterforce campaign against Iraq suggests otherwise. A close look at the likely counterforce targeting objectives and the operational requirements for meeting them

---

[38]Tademy, *Passive Defense & ABO Threat,* Theater Defense Architecture Study, Annex G.

[39]In the strictest terms, "preemptive" counterforce is defined as the use of force to neutralize WMD *already deployed*, while using force to neutralize WMD capabilities still in development is defined as "preventive" counterforce.

reveals a number of critical weaknesses in America's counter-WMD capabilities.

Targeting Objectives. From both a logistical and operational standpoint, counterforce operations designed to neutralize an adversary's WMD programs are likely to be very large in scale, technologically challenging, and extremely complex. And while the size of the WMD target base will determine the ultimate scale of the operation, the need to suppress enemy air defenses rapidly will also drive the attacking force size and composition as well as influence its technical difficulty and overall complexity. In general, counterforce planners will need to consider at least two distinct target sets:

- *WMD-related capabilities.* Including: 1) nuclear targets: nuclear reactors, engineering labs, fuel fabrication facilities, weapon design centers, machine tooling facilities, chemical extraction or enrichment plants, and ammunition depots stockpiling special explosives; 2) biological and chemical sites: research, production and storage facilities for both; and 3) delivery systems: ballistic and cruise missile research, production and storage facilities; mobile missiles and their launchers; and aircraft believed to be equipped for WMD delivery.

- *Air defenses.* These targets include: early-warning radars, both ground-based and airborne (i.e., AWACS-type aircraft); command and control facilities such as sector operation centers (SOCs); communications and computer links and their power sources; surface-to-air missile systems, both fixed and mobile; and enemy aircraft, both in the air and on the ground, as well as their associated airfields and command and control facilities.[40]

Operational Requirements. As this analysis of Iraqi WMD facilities has revealed, the number of targets that may need to be attacked is quite staggering. The size and elusive nature of the overall target base, when combined with the need for rapid action, make the counterforce option operationally daunting.

Accurate and Timely Intelligence. As Ashton Carter, an assistant secretary of defense in the Clinton Administration, puts it, intelligence may be the "long pole in the tent" of any successful counterforce operation. Prior to the initial strikes, planners will need highly accurate intelligence on both the location and physical characteristics of each target. Targets will need to be intensely studied to determine their structural weakpoints. One study, conducted for the United States Air Force, identified five military operations criteria and eighty-one essential elements to evaluate the weapons-target interaction associated with WMD facilities.[41] As shown in this Iraqi case study, some WMD installations will feature numerous structures,

---

[40]Depending on the situation, planners might also consider targeting an adversary's military and/or political leadership and certain segments of its fielded forces. For purposes of illustrating operational requirements, however, it will suffice to focus on only those targets likely to be deemed essential in any counterforce operation. For a more comprehensive look at the range of possible targeting objectives, see Zelikow, "Offensive Military Options," pp. 174–85.

[41]Robert W. Chandler, *Counterproliferation Employment Planning Analysis Tool* (Arlington, Va.: Strategic Planning International, Inc., November 1994).

each of which may be compartmentalized, requiring planners will to identify several different critical aimpoints for each target area.

Throughout the operation, battle damage assessment (BDA), real-time battlefield surveillance, and battle management become equally critical. The Gulf War illustrated the difficulty of assessing damage done in air attacks. Ironically, precision weapon damage is particularly difficult to assess—all that may be visible after a direct hit is a "tiny" hole in a structure's roof.[42] In Desert Storm, however, planners had the luxury of "the next day" to make sure the desired damage had been done. Even then the results were mixed. Some hardened aircraft shelters at Iraqi airfields were hit several times while others just hundreds of yards away were untouched. In future counterforce operations, however, the possibility of WMD retaliation and/or dispersal of remaining WMD assets will place a premium on rapid assessment and re-strike. Finally, there is the problem of mobile targets. In Desert Storm, nearly 1,500 sorties were flown against mobile Scud launchers with no confirmed kills.[43] The success of any future theater-based counterforce operation is likely to hinge on the ability of U.S. forces to track and destroy mobile WMD-armed ballistic missiles rapidly—"truly the war-stoppers of the future."[44] General Richard E. Hawley, then-commander in chief, U.S. Air Forces in Europe, characterized the problem:

---

[42]Robert D. Blackwill and Ashton B. Carter, "The Role of Intelligence," in *New Nuclear Nations: Consequences for U.S. Policy*, ed. by Robert D. Blackwill and Albert Carnesale, p. 241.

[43]U.S., Air Force, *Gulf War Airpower Survey: Operations and Effects and Effectiveness,* Vol. II, Part 2, pp. 330–32.

[44]Lieutenant General Jay M. Garner, Chief, Army Space and Strategic Defense Command, as quoted in Kitfield, "Counterproliferation," p. 57.

The key to hunting mobile missiles is, number one, having a good array of ground, air, and space sensors and then linking them to a command-and-control system that can get the information to a pilot fast enough for him to put ordnance on the target. That's not a trivial problem. In fact, it's very difficult.[45]

Another defense official likened the challenge of "Scud hunting" to anti-submarine warfare: "the major challenge is finding the bastards, [not] killing them."[46] There has been some progress toward developing a viable capability against today's missiles, but those involved in the effort concede that the threat is getting more complex.

Initial Decisiveness. To prevent dispersal of critical WMD assets and possible WMD retaliation, the initial strikes in a counterforce operation must be decisive. While many types of military forces may be required, land- and sea-based airpower will no doubt comprise the main strike force. Planners will want to destroy as many targets as they can as quickly as possible. This places a premium on surprise, lethality and payload.

Surprise is best achieved in air operations with stealth attack platforms that can approach their targets without alerting the enemy. Aircraft range will also be critical to achieving surprise. The closer U.S. aircraft are based to the adversary, the more likely he would be to detect their presence in the region and begin

---

[45]As quoted in Kitfield, "Counterproliferation," p. 57.

[46]Lieutenant Colonel Dan Kirby, Director of Joint Project Optic Cobra, the anti-missile exercise occurring as part of the annual Roving Sands air defense training exercise. As quoted in Kitfield, "Counterproliferation," p. 57.

taking evasive and/or military countermeasures (such as retaliation against U.S. air bases and aircraft carriers). Long-range cruise missiles, though not necessarily "stealthy" in terms of radar signature, can also be useful in achieving surprise. Their small size and terrain-following mission profile make them virtually indistinguishable on current radar screens from other "noise" at their altitude; their long range, meanwhile, facilitates delivery from outside enemy detection zones by non-stealthy bombers, surface ships and submarines.

Lethality involves the weapon-target interaction wherein weapon accuracy, explosiveness, and penetration capability are paramount. Laser-guided weapons are generally considered the most accurate aircraft-delivered munitions, but they cannot be delivered in adverse weather or through heavy smoke. Inertially-guided precision weapons utilizing Global Positioning System (GPS), on the other hand, can be delivered through smoke and in any weather, and are demonstrating accuracies nearly equal to their laser-guided counterparts. Using inertially-guided weapons, therefore, dramatically reduces the potential for critical disruptions caused by bad weather or lingering smoke.[47]

Generally speaking, the heavier the weapon, the greater the damage. Thus, 1000- and 2000-lb precision munitions such as the Joint Direct Attack Munition (JDAM), available in usable numbers by 1999, will be, in most cases, the U.S. weapon of choice against WMD targets. Against hardened underground

---

[47]The Global Positioning System (GPS) consists of 24 satellites orbiting 11,000 miles above the Earth that broadcast signals continually. With the right receivers, many of them commercially available, one can precisely identify locations on the ground and water and in the air to an accuracy of ten meters.

facilities, however, the U.S. will need penetration capabilities not found in current generation non-nuclear weapons. Recognizing this problem, the Air Force plans to attach GPS guidance kits to 4,700-lb BLU-113 bombs, transforming them into heavy penetrating direct-attack munitions. In counterforce operations, these massive weapons would almost certainly need to be delivered by the B-2, the only heavy bomber capable of penetrating the thick defenses expected to be protecting critical WMD targets.

Finally, aircraft payload is critical for the simple reason that many targets must be destroyed in a very short amount of time. The more weapons per sortie, the fewer the sorties required to destroy a target set. The most valuable airpower assets in counterforce operations will therefore be those possessing stealth, long range, large payload capacities, and precision delivery capabilities (preferably all-weather).

Assessing U.S. Counterforce Assets. In reviewing potential U.S. forces for counterforce operations, the following points summarize our observations:

- The B-2 bomber, which offers an unprecedented combination of stealth, GPS-aided all-weather precision, intercontinental range, and large payload will undoubtedly be America's most critical counterforce platform.[48] Properly equipped, B-2s will be capable of

---

[48]The B-2 has a range of > 6,000 nm (unrefueled) and a payload of > 40,000-lb. Each B-2 can carry 84 1,000-lb bombs, 16 2,000-lb inertially-guided JDAMs, or eight 4,700-lb GPS-Aided BLU-113 penetrating munitions. Source: Northrop Grumman Corporation.

destroying every type of WMD-related target, from mobile missiles to hardened bunkers, and can do so from bases beyond retaliatory range without being detected. However, only sixteen aircraft will be available for combat when the fleet reaches full operational capability in 2000. Additional B-2s would be required if America wishes to field a truly decisive long-range precision strike force.[49]

• The F-117 medium attack aircraft, star of Desert Storm, can in some ways be considered a mini-B-2. Stealthy like the B-2, F-117s can also deliver precision munitions through the most sophisticated air defenses. Yet F-117s carry only two 2,000-lb bombs each, need good weather to deliver their laser-guided weapons, and require theater basing, which could compromise operational security and leaves them vulnerable to WMD retaliation. Moreover, there are only thirty-six operational F-117s in the force and production ceased in 1989.

• After stealth aircraft, long-range conventional standoff weapons such as the Air Force's AGM-86 Air-Launched Cruise Missile (ALCM-C) and the Navy's sea-launched BGM-109 Tomahawk Land Attack Missile (TLAM-C) would be the most valuable U.S. counterforce assets. Air Force heavy bombers and Navy submarines and surface

---

[49]For a thorough discussion of sizing the B-2 force for large-scale long-range strike operations, see Glen Buchan, "The Use of Long-Range Bombers in a Changing World: A Classical Exercise in Systems Analysis," in *New Challenges for Defense Planning: Rethinking How Much Is Enough*, ed. by Paul K. Davis (Santa Monica: RAND, 1994), pp. 432–33.

ships are the primary cruise missile delivery platforms. Both the ALCM-C and the TLAM-C possess excellent range, and both are combat-proven to be accurate and highly effective at penetrating air defenses.[50] However, like all current generation standoff weapons, both cruise missile variants lack the punch required to destroy hardened and/or buried targets. And, because they require preprogramming, they cannot be targeted against mobile missile launchers or WMD assets relocated by the adversary after the initial strikes.

- The B-52H and B-1B bombers offer long range and large payload; like the B-2, these aircraft can strike from distant bases. Their lack of stealth, however, limits their role to delivering long-range standoff weapons when facing heavy air defenses.[51] A B-52H can carry twenty cruise missiles; B-1Bs can carry eight (though such a configuration raises arms control issues). Standoff

---

[50]The ALCM-C has a range of over 1,500 nm; TLAM-C range is listed at over 700 nm. Source: "Missile Forecast," *Forecast International/DMS Market Intelligence Report*, 1995.

[51]There are a number of shorter- and medium-range standoff weapons planned, including the AGM-84E Standoff Land-Attack Missile (SLAM; 120 mi. range), the AGM-142 HAVE NAP missile (50 mi. range), the AGM-154 Joint Standoff Weapon (JSOW; 40 mi. range), and the Joint Air-to-Surface Standoff Missile (JASSM; 180 mi. range). However, these weapons must, because of their limited range, be launched from within reach of most enemy early-warning radars and air interceptors, limiting their utility for non-stealthy aircraft in high-threat environments. Source for missile ranges: Teal Group Corporation, *World Missiles Briefing* (March 1995). See also Glen C. Buchan and David R. Frelinger, *Providing an Effective Bomber Force for the Future*, CT-119 (Santa Monica: RAND, May 1994), pp. 4–9.

weapons, however, are not nearly as efficient in the use of heavy bomber payload volumes as direct-attack weapons such as JDAM.

- Given the dearth of highly-capable long-range aircraft, land-based F-15E and F-16C/D and carrier-based F/A-18 fighters would be expected to carry the bulk of the strike load in any near-term counterforce operation. All are capable of delivering precision munitions (mostly laser-guided). None of these fighters, however, are stealthy, their payloads are limited, and all require theater basing.

Considering that the number of aimpoints in a counterforce attack could range well into the thousands, that some of the aimpoints may require several weapons, and that both penetrating bombers and cruise missiles will be in short supply, the U.S. would need several hundred land- and sea-based strike fighters in-theater at the start of the operation.[52] Supporting the strike aircraft would be large numbers of air escort, air defense suppression and radar jamming aircraft (which in the Gulf War's

---

[52]Because of their high unit cost (<$1 million), Air Force and Navy cruise missile buys have been limited, as has cruise missile usage in U.S. combat operations. During the Gulf War, for example, cruise missile usage was terminated relatively early on "due to high cost concerns relative to the use of reusable stealth assets." See U.S., Commission on Roles and Missions of the Armed Forces, *Future Bomber Force*, Staff Study (June 1995), p. 14. Sixty-four percent of the 282 TLAM-Cs used in Desert Storm were launched during the first two days of the air campaign, and none were launched after February 1, 1991, with twenty-eight days (65 percent) of the Gulf War remaining. All thirty-five ALCM-C launches took place on the war's first night. See U.S., Department of the Air Force, *Gulf War Airpower Survey: Summary Report*, pp. 200–24.

opening blows comprised almost half the sorties flown).[53] Many tankers would also be needed to support the short-range fighters. The total number of aircraft participating in the operation will vary with the size of the WMD target base, but one can assume that it would number many hundreds of aircraft and most would require theater basing.[54]

America's current and projected reliance upon short-range tactical aircraft for precision strike bodes ill for the success of future counterforce operations, especially surprise "preemptive" strikes. It is difficult if not impossible to imagine the United States deploying hundreds of fighters and multiple carrier battle groups into a region without tipping its hand to the target country. Surprise would be lost and the adversary could disperse many critical WMD assets, reducing the chances of success. Regional allies, whose bases and facilities would be of the utmost importance, could very well face WMD threats from the target country and might balk at the notion of supporting the U.S. deployment. Should these countries agree to cooperate, there would be tremendous pressure on the U.S. to not only succeed, but possibly to expand its operational objectives and to guarantee

---

[53]Extracted from unclassified portions of the "Master Attack Plan: First 24 Hours," printed at 21:21 hours Riyadh time, January 16, 1991. According to the plan's author, then-Lt. Col. David A. Deptula, "the master attack plan was a new planning document developed . . . specifically to facilitate planning the Gulf War air campaign." See Colonel David A. Deptula, *Firing for Effect: Change in the Nature of Warfare* (Arlington, Va.: Aerospace Education Foundation, August, 1995), p. 23.

[54]Bombers, tankers, and possibly surveillance and C³I aircraft can stage operations from outside the theater.

rapid military reinforcement should regional hostilities erupt.[55] Finally, all theater-based (including carrier-based) aircraft within striking range of their targets would be highly vulnerable to the very WMD they were sent to destroy. Indeed, such an operation would be tantamount to full-scale war with a WMD-armed adversary—literally a self-fulfilling prophecy if the intent of the operation was to preempt WMD use against U.S. forces.

Preemption. Despite being officially disavowed by the Clinton Administration, the concept of preemptive counterforce is nonetheless attracting the interest of many in the national security community. Preemptive counterforce appears to offer a promising solution to a very difficult problem: as WMD threats emerge, the U.S. can simply "neutralize" them in the pre-war time frame before facing them in battle. The reality of such operations, however, suggests otherwise. Considering the massive scale of such operations, and since the target country would likely be prepared for the attack, it is difficult to consider the option truly "preemptive." In such operations, the U.S. would clearly be inviting WMD use against its own forces or its regional friends and allies, thus defeating the purpose of the operation. It therefore appears unlikely the U.S. would resort to counterforce against a mature WMD program except in the context of a greater regional conflict, where such operations would indeed become a top priority.

This brings us to the preemption paradox: the United States is more likely to want to attack WMD programs it cannot confidently neutralize and less likely to want to attack the ones it

---

[55]Zelikow, "Offensive Military Operations," p. 186.

can. As former National Security Council staff member Philip Zelikow notes, an "inverse relationship" exists between the threat posed by proliferators and their vulnerability to offensive military action.[56] In general, the more mature the WMD program, the greater the threat, and the less likely counterforce will succeed—and vice versa. The reasons are fairly simple to understand. Mature programs are more likely to resemble Iraq circa January 1991: large, dispersed, relocatable, hardened, potentially weaponized, and heavily defended. In immature WMD programs, meanwhile, there are comparatively fewer critical facilities; installations—are more likely to be geographically centralized and fixed; WMD assets are not likely be weaponized; and targets are likely to be less heavily-defended.

Yet, as Zelikow points out, immature programs are not likely to be the target of future U.S. counterforce operations. During the early stages of development, the United States will have difficulties discerning the true purpose of a WMD program and thus the nature of the threat. Absent a clear threat, it is highly unlikely that the United States would resort to preemptive strikes. In cases where the proliferator appears to be on the threshold of fielding operational WMD, there may be support from many quarters for preemption, and their may also be a limited window of opportunity for successfully neutralizing the emerging threat.[57] But there will also be fierce opposition—both domestic and international—to the forcible disarmament of another country

---

[56]*Ibid.*, pp. 164–65.

[57]*Ibid.*, p. 168 (Zelikow defines a "limited period" as "several months, perhaps even a year or two.").

with which the U.S. is not already at war.   As Michele Flournoy, currently a deputy assistant secretary of defense, has noted:

> A preventive war in the absence of a larger crisis would be less likely to receive [domestic] popular support and would raise some profoundly unsettling moral questions: Would the United States be justified in initiating hostilities for the sole purpose of denying an adversary a nuclear weapons capability?   Would the amount of force required to accomplish this, and the damage it could cause, be proportionate to the objectives involved?[58]

And, while the international community generally favors reversing proliferation wherever possible, the initial foreign response to the Clinton Administration's Counterproliferation Initiative illustrates that there is no such consensus on using force to do so.   To be sure:

> Any preventive war is. . . likely to be highly controversial. Even in cases where the proliferator is widely seen as a pariah, such as North Korea or Libya, building a multilateral coalition to support a preventive war could prove difficult, if not impossible . . . U.S. decisionmakers [must] think past the Gulf War experience of preventive strikes in the context of a war already begun, and the tremendous international support they received, to the far more difficult dilemma of determining whether to start a preventive war in peacetime.   They must try to imagine

---

[58]Flournoy uses the terms "preventive" and "preemptive" interchangeably.   See Flournoy, "Implications for U.S. Military Strategy," pp. 149–50.

how the U.S. bombing if Iraqi [WMD] facilities might
have been received by the international community if Iraq
had not invaded Kuwait.[59]

The bottom line here is that counterforce operations can no
longer be considered in a vacuum, separate from mainstream U.S.
military strategy. The size, scale and potential political
ramifications of any future counterforce operation are more likely
to resemble the opening days of Operation Desert Storm than
those of America's 1986 bombing raid against Libya. Indeed,
preemptive counterforce operations against proliferators who
most threaten the vital interests of the United States can only be
thought of as preemptive *war*. And, since it is during the course
of regional war that the U.S. is most likely to feel the sting of
WMD, a war by any name is an option the U.S. will be least
likely to choose.

If counterforce operations are to be considered a viable option
for countering the regional WMD threat under the condition of
inadequate theater missile and air defenses, the United States
must dramatically reduce the size of its theater "footprint." That
is, the U.S. must extend its own "theater of operations" beyond
the range of enemy's WMD without compromising operational
effectiveness.

---

[59]*Ibid.*, p. 150.

Conclusions: The Road Ahead

Under the present budget-driven U.S. military strategy, WMD-armed regional aggressors will retain the ability to critically limit America's power projection capabilities. Despite the comprehensive plan for countering proliferation that the current Administration inherited from the Bush Administration and the intellectual drive of Les Aspin that propelled the Counterproliferation Initiative, the three-pronged approach for countering the wide-ranging regional WMD threat has been unhinged. *Deterrence* can never be guaranteed. *Active defenses* have been fiscally stretched and *passive defenses* amount to little more than damage control. *Counterforce* operations will remain militarily infeasible so long as theater-based aircraft are left to carry the precision strike burden, and preemptive operations appear to be a non-starter. There is, in fact, little the United States can do to negate the many implications of proliferation short of "extending" the theater of operations beyond the range of enemy WMD.

The United States is, however, capable of developing a robust, if imperfect, 21st century counterforce capability. By heavily augmenting its long-range precision striking power, particularly the heavy bomber and long-range cruise missile forces, the U.S. can credibly threaten to exact a heavy toll on aggressor WMD programs worldwide—without subjecting itself to the litany of problems faced by a largely theater-based force. Operational effectiveness will continue to be limited to some degree by various technological challenges, such as those associated with intelligence-gathering and hard-target penetration. "Effectiveness" here is a relative concept. In the eyes of a potential proliferator, the damage potential of a beefed-up and largely invulnerable long-range strike force might discourage WMD development and/or acquisition. In conflicts with known proliferators, such a capability would allow the U.S. conduct relatively risk-free counterforce strikes prior to large-scale, and otherwise highly vulnerable, theater force deployments. Absent the immediate threat of theater WMD retaliation, long-range counterforce operations could be protracted, allowing the U.S. to sustain strikes until it is deemed "safe" to enter the theater.

But this begs a larger, more fundamental, and far more important question: If the United States has within its technological and fiscal grasp the capability to strike decisively from outside the range of WMD, why doesn't the U.S. shift the balance of its military posture in this direction? Why should the core military strategy still entail sending sizable numbers of American forces into a proliferator's backyard, where the success of the strategy hinges upon our capability to deter, defend against, or defeat his WMD, none of which can ever be guaranteed?

The real question is not how can the United States deter WMD use, neutralize enemy WMD capabilities, or defend against WMD attack. The real question is how can the United States deter the regional aggression that WMD are designed to support? The answer: by *demonstrating* the capability to stop that aggression with forces operating from beyond WMD range—forces invulnerable to theater WMD.[60] Should deterrence

---

[60]For a thorough discussion of the strategic and military requirements for deterring regional adversaries, see Wilkening and Watman, *U.S. Regional Deterrence Strategies*, especially pages xii–xiv, 81–82, and 85–86. In short, Wilkening and Watman argue that deterrence of the most "highly motivated"

fail, the U.S. would then feel confident that it can take swift and decisive military measures to protect its threatened interests.

This is not to suggest that the United States should simply scrap its tactical, or short-range strike components. Far from it. We are, however, recommending that more of the warfighting *burden* be shifted from forces dependent upon regional basing to those that can operate effectively from outside the region of conflict. The first phase of any U.S. attack or counterattack is likely to be an air campaign. Air-delivered weapons can be used against strategic targets and/or advancing armies. Why should such a campaign not be staged from beyond the range of the weapons that can really hurt the U.S. expeditionary forces or cause American political and military leaders to balk at taking decisive action in the first place?

Long-range precision strike, even against mobile systems such as advancing armor and mobile missile launchers, is technologically possible now. Stealth bombers, capable of operating without theater support, can now accomplish tasks once reserved exclusively for theater-based fighter aircraft.[61] Long-

---

regional aggressors will require nothing short of a "prompt denial" capability—a *demonstrated* capability to prevent potential adversaries from reaching their initial military objectives. According the authors, the military forces required for a prompt denial capability are forward-deployed forces or those that are so rapidly deployable as to be "virtually" stationed in the theater. Slower-deploying U.S. conventional forces (i.e., theater-range forces) are "less relevant for deterrence" because regional adversaries "often do not believe such forces will arrive" in time to stop them.

[61]By "tasks" we are referring to precision ground-attack missions; bombers do not currently possess an air-to-air capability. Stealthy bombers can, however, greatly contribute to the achievement of air superiority through offensive

range cruise missiles, though less flexible than penetrating aircraft, offer many important advantages and will increase in capability through technological advances. Neither are vulnerable to theater-range WMD. The B-2 is clearly the ultimate manned strike platform for a WMD world—both in a counterforce and general warfighting sense—but the Clinton Administration and Pentagon officials argue that we should spend future resources on new theater-range fighters (F-22, F-18E/F, and Joint Strike Fighter), not bombers. The Air Force does not have a long-range cruise missile in production or development. The Navy plans an inventory of less than 4,000 TLAMs. Simply put, the United States is passing on what is clearly the best defense against weapons of mass destruction: being somewhere else when they go off.

As long as the United States continues to subscribe to the Bottom-Up Review strategy of pouring hundreds of thousands of forces within a proliferator's reach, any and all counterproliferation measures will be tantamount to "tinkering at the margins." This can lead to only one conclusion: the strategy is flawed. The BUR also drives the Defense Department's force modernization plans which, according to Pentagon officials, is too tightly-woven to accommodate a bomber force expansion. But if the "plan" does not allow the United States to exploit its strengths and field the most appropriate fighting force possible, the plan is wrong and a new blueprint is in order.

---

counter-air operations. Moreover, they are inherently capable of achieving one of the primary air superiority objectives: the ability to conduct offensive air operations with impunity.

In his 1994 counterproliferation progress report to Congress, Deputy Defense Secretary John Deutch said that it is "not easy to change the direction of the ship of the state—especially when its course for over 45 years was primarily aimed at preparing for threats that have receded, while the problems of proliferation have grown and become more urgent."[62]   It is, therefore, hardly surprising that the regional WMD threat is being considered in a vacuum, largely separate from mainstream military planning. Hamstrung by a budget-driven strategy-making process, the U.S. military establishment has had an exceedingly difficult time seeing the big picture as it struggles to adapt to the post-Cold War era. Yet, this cannot continue if the United States is to effectively counter the regional WMD threat. Military strategy must be designed to satisfy U.S. national security objectives.

The proliferation of weapons of mass destruction must be recognized as a fundamental change in the security environment, and the implications of WMD-armed adversaries must be accounted for at every stage of the military planning process. Unfortunately, the Pentagon, despite recent progress, still seems determined to plan for the wrong war and the wrong threat. The potential outcome in a future conflict could well be disastrous.

---

[62]Deutch, *Report on Nonproliferation and Counterproliferation Activities and Programs*, p. 25.

Selected Bibliography

Books, Journals, and Periodicals:

Albright, David and Hibbs, Mark. "Iraq and the Bomb: Were They Even Close?" *Bulletin of the Atomic Scientists* (March 1991).

―――. "Iraq's Nuclear Hide-And-Seek." *Bulletin of the Atomic Scientists* (September 1991).

Bailey, Kathleen C. *The UN Inspections in Iraq: Lessons for On-Site Verification.* Boulder: Westview Press, 1995.

Ballistic Missile Proliferation: An Emerging Threat. Arlington, Va.: System Planning Corporation, 1992.

Blackwill, Robert D. and Carter, Ashton B. "The Role of Intelligence." *New Nuclear Nations: Consequences for U.S. Policy.* Edited by Robert D. Blackwill and Albert Carnesale. New York: Council on Foreign Relations, 1993.

Bruce, James and Starr, Barbara. "US Exploits Images of Military Rebirth... As Iraq Rejects UN Resolution on Oil Sales." *Jane's Defence Weekly* (May 6, 1995).

Bruce, James. "Playing Hide and Seek With Saddam." *Jane's Defence Weekly* (January 6, 1996).

Buchan, Glen. "The Use of Long-Range Bombers in a Changing World: A Classical Exercise in Systems Analysis." *New Challenges for Defense Planning: Rethinking How Much Is Enough.* Edited by Paul K. Davis. Santa Monica: RAND, 1994.

Carnesale, Albert. "Defenses Against New Nuclear Threats." *New Nuclear Nations: Consequences for U.S. Policy*. Edited by Robert D. Blackwill and Albert Carnesale. New York: Council on Foreign Relations, 1993.

Chandler, Robert W. *Counterproliferation Employment Planning Analysis Tool*. Arlington, Va.: Strategic Planning International, Inc., November 1994.

Crossette, Barbara. "Experts Doubt Iraq's Claims On A-Bomb." *New York Times* (August 30, 1995).

Deen, Thalif. "UNSCOM Accuses Iraq of Purchasing Arms." *Jane's Defence Weekly* (October 21, 1995).

Deptula, David A. Colonel. *Firing for Effect: Change in the Nature of Warfare*. Arlington, Va.: Aerospace Education Foundation, August 1995.

Dunn, Lewis. "New Nuclear Threats to U.S. Security." *New Nuclear Nations: Consequences for U.S. Policy*. Edited by Robert D. Blackwill and Albert Carnesale. New York: Council on Foreign Relations, 1993.

Ebert, Barbara. "Iraq: Its Nuclear Past As a Way of Assessing Its Nuclear Future." Unpublished report. Vienna, Va.: Science Applications International Corporation, April 1994.

Edensword, Diana and Milhollin, Gary. "Iraq's Bomb--An Update." *New York Times* (April 26, 1993).

Eisenstadt, Michael. *Like a Phoenix From the Ashes? The Future of Iraqi Military Power*. Policy Paper No. 36. Washington, D.C.: Washington Institute for Near East Policy, 1993.

————. *"The Sword of the Arabs:"* *Iraq's Strategic Weapons.* Policy Paper No. 21. Washington, D.C.: Washington Institute for Near East Policy, 1990.

Flournoy, Michele A. "Implications for U.S. Military Strategy." *New Nuclear Nations: Consequences for U.S. Policy.* Edited by Robert D. Blackwill and Albert Carnesale. New York: Council on Foreign Relations, 1993.

Garrity, Patrick. "Implications of the Persian Gulf War for Regional Powers." *Weapons Proliferation in the 1990s.* Edited by Brad Roberts. Cambridge, Mass: MIT Press, 1995.

Gordon, Michael R. and Trainor, Bernard E., General. *The Generals' War: The Inside Story of the Conflict in the Gulf.* Boston: Little, Brown & Company, 1995.

Hitchens, Theresa. "Wargame Finds U.S. Falls Short In Bio War." *Defense News* (September 3, 1995).

————. "U.S. Lawmakers Want Tighter Lid on Nuke Material." *Defense News* (March 18–24, 1996).

Infield, Glen B. *Disaster at Bari.* New York: Macmillan, 1971.

"Iraq Finally Admits Building Biological Weapon Arsenal." *Jane's Defence Weekly* (July 15, 1995).

Isby, David C. "The Residual Iraqi 'Scud' Force." *Jane's Intelligence Review,* 7–3 (March 1995).

Jane's Intelligence Review POINTER (September 1995).

Kay, David A. "Denial and Deception Practices of WMD Proliferators: Iraq and Beyond." *Weapons Proliferation in the 1990s*. Edited by Brad Roberts. Cambridge, Mass: MIT Press, 1995.

"King Hussein Denounces Killing of Repatriates." *Washington Post* (February 26, 1996).

Kitfield, James. "Counterproliferation." *Air Force Magazine* (October 1995).

Latter, Richard Dr. "The Increased Danger of Biological Weapons Proliferation." *Jane's Intelligence Review* (February 1994).

Luttwak, Edward N. "Toward Post-Heroic Warfare." *Foreign Affairs* (May/June 1995).

Mandel, Robert. "Chemical Warfare: Act of Intimidation or Desperation?" *Armed Forces & Society* (Winter 1993).

Millot, Marc, Molander, Roger, and Wilson, Peter. *"The Day After . . ." Study: Nuclear Proliferation in the Post-Cold War World.* Vol. I. Summary Report MR-266-AF (Santa Monica: RAND, 1993).

"Missile Forecast." *Forecast International/DMS Market Intelligence Report*, 1995.

Moodie, Michael. "Beyond Proliferation: The Challenge of Technology Diffusion." *Washington Quarterly* (Spring 1995).

Müller, Harald and Reiss, Mitchell. "Counterproliferation: Putting New Wine in Old Bottles." *Washington Quarterly* (Spring 1995).

Nerlich, Uwe. *Toward a New Nuclear Weapons Regime.* Prepared for Sandia National Laboratories. Ebenhausen, Germany: Stiftung Wissenshaft und Politik, Forschungsinstitut für Internationale Politik und Sicherheit, August 1993.

"Nuclear Watchdog Unleashes New Powers to Stall Evaders." *Jane's Defence Weekly* (April 8, 1995).

Porteus, Holley. "Grappling With the BW Genie." *International Defense Review* (March 1995).

Powers, John R. and Muckerman, Joseph E. "Rethink the Nuclear Threat." *Orbis* (Winter 1994).

Rathmell, Andrew, Dr. "Chemical Weapons in the Middle East: Lessons From Iraq." *Jane's Intelligence Review* (December 1995).

"Riding High: Corporate America Now Has An Edge Over Its Global Rivals." *Business Week* (October 9, 1995).

"Saddam Hopes BW Confession Is Enough to Convince USA." *Jane's Defence Weekly* (September 2, 1995).

"Saddam's Secret Bomb." London ITV Television Network (1830 GMT, May 19, 1994). U.S. Foreign Broadcast Information Service, JPRS Report. *Proliferation Issues.* JPRS-TND-94–013 (June 24, 1994).

Smith, R. Jeffrey and Frankel, Glenn. "Saddam's Nuclear-Weapons Dream: A Lingering Nightmare," *Washington Post* (October 13, 1991).

Smith, R. Jeffrey. "Iraq Buying Missile Parts Covertly." *Washington Post* (October 14, 1995).

————. "Iraq Had Program For Germ Warfare." *Washington Post* (July 6, 1995).

————. "Iraq Is Hiding 6 to 16 Scuds, U.N. Suspects." *Washington Post* (March 21, 1996).

————. "Iraq's Nuclear Prowess Underestimated by U.S." *Washington Post* (October 13, 1991).

————. "U.N. Says Iraqis Prepared Germ Weapons in Gulf War." *Washington Post* (August 26, 1995).

Teal Group Corporation, *World Missiles Briefing* (March 1995).

Teischer, Howard. "The Naive Hope That Allowed Hussein to Weigh Mass Murder." *Los Angeles Times* (September 10, 1995).

Toffler, Alvin and Heidi. *War and Anti-War*. New York: Little, Brown, 1993.

"U.N. Official: Iraq Worked On Radiological Arms." *Washington Post* (November 8, 1995).

Weiner, Tim. "U.S. Vulnerable to Terrorist Chemical Weapons." *New York Times* (March 21, 1996).

Wilkening, Dean and Watman, Kenneth. *Nuclear Deterrence in a Regional Context*. Santa Monica: RAND, 1994.

Williams, Phil and Woessner, Paul. "The Real Threat of Nuclear Smuggling." *Scientific American* (January 1996).

Woolsey, R. James. "Missile Blackmail Exists Today." *Wall Street Journal* (April 8, 1996).

Zelikow, Philip. "Offensive Military Options." *New Nuclear Nations: Consequences for U.S. Policy.* Edited by Robert D. Blackwill and Albert Carnesale. New York: Council on Foreign Relations, 1993.

Zimmerman, Peter D. "Proliferation: Bronze Medal Technology Is Enough." *Orbis* (Winter 1994).

Public Documents:

International Atomic Energy Agency:

International Atomic Energy Agency. "Fact Sheet" (January 1994). Enclosed in a letter to the author from Maurizio Zifferero, Leader of UNSC 687 Action Team. Vienna, Austria: International Atomic Energy Agency, January 18, 1994.

————. *IAEA Inspections and Iraq's Nuclear Capabilities.* Vienna, Austria: April 1992.

Lopez Lizana, Fernando, Ouvard, Robert, and Takats, Ferenc. "Nuclear Inspections in Iraq: Removing Final Stocks of Irradiated Fuel." *IAEA Bulletin*, 3/1994.

United Nations:

United Nations, Security Council. *Consolidated Report on the First Two IAEA Inspections Under Security Council Resolution 687 (1991) of Iraqi Nuclear Capabilities* (July 11, 1991). U.N. Doc. No. S/22788 (July 15, 1991).

————. *Eighth Report of the Executive Chairman of the Special Commission Established By the Secretary-General Pursuant to Paragraph 9 (b) (i) of Security Council Resolution 687 (1991), On the Activities of the Special Commission.* U.N. Doc. S/1994/1422, Appendix I (December 15, 1994).

————. *First Report on the Sixth IAEA On-Site Inspection in Iraq Under Security Council Resolution 687 (1991)* (September 22–30, 1991). U.N. Doc. S/23122 (October 8, 1991).

————. *Fourth Report of the Executive Chairman of the Special Commission Established By the Secretary-General Pursuant to Paragraph 9 (b) (i) of Security Council Resolution 687 (1991), On the Activities of the Special Commission.* U.N. Doc. S/24984 (December 17, 1992).

————. *Ninth Report of the Executive Chairman of the Special Commission Established By the Secretary-General Pursuant to Paragraph 9 (b) (i) of Security Council Resolution 687 (1991), On the Activities of the Special Commission.* U.N. Doc. S/1995/494 (June 20, 1995).

————. *Report of the Eighth IAEA On-Site Inspection in Iraq Under Security Council Resolution 687 (1991)* (November 11-18, 1991). U.N. Doc. S/23283 (December 12, 1991).

————. *Report of the Executive Chairman of the Special Commission Established By the Secretary-General Pursuant to Paragraph 9 (b) (i) of Security Council Resolution 687 (1991).* U.N. Doc. S/23165 (October 25, 1991).

————. *Report of the Secretary-General on the Status of the Implementation of the Special Commission's Plan for the Ongoing Monitoring and Verification of Iraq's Compliance With Relevant Parts of Section C of Security Council Resolution 687 (1991).* U.N. Doc. S/1995/284 (April 10, 1995).

————. *Report of the Secretary-General on the Status of the Implementation of the Special Commission's Plan for the Ongoing Monitoring and Verification of Iraq's Compliance with Relevant Parts of Section C of Security Council Resolution 687 (1991).* U.N. Doc. S/1995/864 (October 11, 1995).

————. *Report on the Fourth IAEA On-Site Inspection in Iraq Under Security Council Resolution 687 (1991)* (July 27-August 10, 1991). U.N. Doc. S/22986 (August 28, 1991).

————. *Report on the Seventeenth IAEA On-Site Inspection in Iraq Under Security Council Resolution 687 (1991)* (January 25–31, 1993). U.N. Doc. S/25411 (March 13, 1993).

————. *Report on the Seventh IAEA On-Site Inspection in Iraq Under Security Council Resolution 687 (1991)* (October 11–22, 1991). U.N. Doc. S/23215 (November 14, 1991).

————. *Report on the Status of Compliance By Iraq With the Obligations Placed Upon It Under Section C of Security Council Resolution 687 (1991) and Resolutions 707 (1991) and 715 (1991).* U.N. Doc. S/23993 (May 22, 1992).

————. *Report on the Twelfth IAEA On-Site Inspection in Iraq Under Security Council Resolution 687 (1991)* (May 26-June 4, 1992). U.N. Doc. S/24223 (July 2, 1992).

————. *Report on the Twenty-Eighth IAEA On-Site Inspection in Iraq Under Security Council Resolution 687 (1991).* U.N. Doc. S/1995/1003 (December 1, 1995).

————. *Report of the Twenty-Seventh IAEA On-Site Inspection in Iraq Under Security Council Resolution 687 (1991).* U.N. Doc. S/1994/1443 (December 22, 1994).

————. *Resolution 687* (1991). U.N. Doc. S/RES/687 (1991) (April 3, 1991).

————. *Second Report By the Executive Chairman of the Special Commission Established By the Secretary-General Pursuant to Paragraph 9 (b) (i) of Security Council Resolution 687 (1991)*. U.N. Doc. S/23268 (December 4, 1991).

————. *Seventh Report of the Executive Chairman of the Special Commission Established By the Secretary-General Pursuant to Paragraph 9 (b) (i) of Security Council Resolution 687 (1991), On the Activities of the Special Commission*. U.N. Doc. S/1994/750 (June 24, 1994).

————. Special Commission on Iraq. Unpublished information paper (October 16, 1995).

————. *Third Report By the Executive Chairman of the Special Commission Established By the Secretary-General Pursuant to Paragraph 9 (b) (i) of Security Council Resolution 687 (1991)*. U.N. Doc. S/24108 (June 16, 1992).

United States:

Aspin, Les. Congressman. "From Deterrence to Denuking: A New Nuclear Policy for the 1990s." See U.S. Congress. House of Representatives. Committee on Armed Services. Defense Policy Panel. *Shaping Nuclear Policy for the 1990s: A Compendium of Views*. 102d Cong. 2d sess. Washington, D.C.: Government Printing Office, 1993.

————. Secretary of Defense. *Report on the Bottom-Up Review*. Washington, D.C.: Department of Defense, October 1993.

————. Secretary of Defense. "The Defense Department's New Nuclear Counterproliferation Initiative." Address to the National Academy of Sciences. Washington, D.C.: December 7, 1993.

Collins, John M., Davis, Zachary S. and Bowman Steven R. *Nuclear, Biological, and Chemical Weapon Proliferation: Potential Military Countermeasures.* Congressional Research Service. Washington, D.C.: Library of Congress, June 28, 1994.

Collins, John M. *Weapons of Mass Destruction: The Impact of Proliferation on U.S. Military Posture.* Congressional Research Service. Washington, D.C.: Library of Congress, June 2, 1995.

Deutch, John M. Deputy Secretary of Defense. *Report on Nonproliferation and Counterproliferation Activities and Programs.* Washington, D.C.: Government Printing Office, May 1994.

Gold, Ted and Welch, Jasper. Co-Chairmen. *Principal Observations and Context.* Theater Defense Architecture Study. Vol. I. Washington, D.C.: Department of Defense, January 1993.

Perry, William J. Secretary of Defense. *Annual Report to the President and the Congress.* Washington, D.C.: Government Printing Office, 1996.

Tademy, Dudley. *Passive Defense & ABO Threat.* Theater Defense Architecture Study. Annex G. Washington, D.C.: Department of Defense, January 1993.

U.S. Arms Control and Disarmament Agency. "Materials Used in the Manufacture of Chemical and Biological Weapons." *Fact Sheet.* Washington. D.C.: Office of Public Information, October 25, 1993.

————. Arms Control and Disarmament Agency. "Multilateral Nuclear Export Control Regimes." *Fact Sheet.* Washington, D.C.: Office of Public Information, January 24, 1994.

————. Arms Control and Disarmament Agency. "The Missile Technology Control Regime." *Fact Sheet.* Washington, D.C.: Office of Public Information, November 7, 1994.

————. Commission on Roles and Missions of the Armed Forces. *Future Bomber Force.* Staff Study (June 1995).

————. Congress. House of Representatives. Committee on Armed Services. *Intelligence Successes and Failures in Operation Desert Shield/Storm.* 103d Cong. 1st sess. Washington, D.C.: Government Printing Office, 1993.

————. Congress. House of Representatives. Joint Hearing of the House Military Research and Development and Military Procurement Subcommittees, March 7, 1996.

————. Congress. Office of Technology Assessment. *Proliferation of Weapons of Mass Destruction: Assessing the Risks.* Washington, D.C.: Government Printing Office, August 1993.

————. Congress. Office of Technology Assessment. *Technologies Underlying Weapons of Mass Destruction.* Washington, D.C.: Government Printing Office, 1993.

————. Department of Defense. *Conduct of the Persian Gulf War.* Washington, D.C. Government Printing Office, April 1992.

————. Department of State. *Dispatch.* 6-42. October 16, 1995.

————. Department of State. *Dispatch.* 6-43. October 23, 1995.

————. Department of the Air Force. *Gulf War Air Power Survey: A Statistical Compendium and Chronology.* Vol. V. Washington, D.C.: Government Printing Office, 1993.

————. Department of the Air Force. *Gulf War Air Power Survey: Operations and Effects and Effectiveness.* Vol. II. Washington, D.C.: Government Printing Office, 1993.

————. Department of the Air Force. *Gulf War Air Power Survey: Planning and Command and Control.* Vol. I. Washington, D.C.: Government Printing Office, 1993.

————. Department of the Air Force. *Gulf War Air Power Survey: Summary Report.* Washington, D.C.: Government Printing Office, 1993.

————. Foreign Broadcast Information Service. JPRS Report. *Proliferation Issues* (February 5, 1993).

————. Foreign Broadcast Information Service. JPRS Report. *Proliferation Issues* (March 22, 1993).

————. Executive Office of the President. *Annual Report On the Proliferation of Missiles and Essential Components of Nuclear, Biological, and Chemical Weapons.* March 1993.

————. General Accounting Office. *Arms Control: U.S. and International Efforts to Ban Biological Weapons.* GAO/NSIAD-93-113 (December 1992).

————. General Accounting Office. *U.S. Military Items Exported or Transferred to Iraq in the 1980s.* GAO/NSIAD-94-98 (February 1994).

———. Senate. Committee on Governmental Affairs. *Proliferation in the 1990's*. 103d Cong. 1st Sess. Washington, D.C.: Government Printing Office, March 24, 1993.

Index

Ababil-100, 124. *See also* missiles, ballistic and cruise, Iraq.
Abu Ghurayb, 68, 80
Abu Sukhayr Mine, 35
Active defenses, 175-81, 200
Aflatoxin, 74, 76, 77, 78, 81. *See also* biological warfare agents.
Air-breathers, aircraft and cruise missiles, 157, 175
Air defenses, 155, 175, 176-78, 186, 189n, 192, 193, 194, 199
Air Force, U.S.
Bomb damage assessment, 19, 155
Counter-missile operations, 127
Effectiveness of air campaign, 89
GPS guidance kits for bombs, 198
Gulf War Airpower Survey, 19
Long-range cruise missile, 203
Operations-intelligence interface, 186
Air Launched Cruise Missile, Conventional (ALCM-C, AGM-86), 193, 194n
Airborne Warning and Control System (AWACS), 186
Akashat, 86
Al Abbas, 105. *See also* missiles, ballistic and cruise, Iraq.

Al Abid, 105. *See also* missiles, ballistic and cruise, Iraq.
Al Ameer, 48
Al Amir, 48
Al Atheer, 48-49, 53
Albright, David, 21
Al Fahd 300, 105. *See also* missiles, ballistic and cruise, Iraq.
Al Fahd 500, 105. *See also* missiles, ballistic and cruise, Iraq.
Al Fallujah, 86, 87, 88, 91
Al Fallujah I, 91
Al Fallujah II, 92
Al Fallujah III, 92
Al Furat, 41, 45
Algeria, 126
Al Hadre, 48
Al Hakam, 71-72, 74, 78, 79
Al Hijarah, 105. *See also* missiles, ballistic and cruise, Iraq.
Al Hussein, 90, 104, 105, 124, 125. *See also* missiles, ballistic and cruise, Iraq.
Al Jesira, 35, 39, 41, 48
Al Kinde, 68, 78, 80, 124
Al Latifiyah, 68
Al Muhammediyat, 91
Al Muthanna, 74, 85, 91, 92, 93
Al Muthanna State Establishment, 91, 93
Al Qaim, 16, 39
Al Qa qaa, 48

Al Rabiyah Manufacturing Plant, 47
Al Radwan, 48, 140
Al Taji, 109
Al Tuwaitha, 16, 22, 25, 29, 30, 33, 35, 39, 41, 48, 53, 57
Anthrax, 63, 64, 68, 70, 71, 74, 78, 80. *See also* biological warfare agents.
Arab world, 169
Arms Control and Disarmament Agency, 150
Argentina, 104
Ash Sharqat, 39, 45, 48
Aspin, Les (the late), 165, 166, 167, 168, 169, 175, 200
Atomic Vapor Laser-Isotope Separation (AVLIS), 29
Australia Group, 6n, 65, 85, 133, 135
Austria, 104
Axis Powers, 161n

B-52H bombers, 193
B-1B bombers, 192, 193
B-2 bombers, 191-92, 203
BADR 2000, 104. *See also* missiles, ballistic and cruise, Iraq.
BLU-113 bombs, GPS-aided, 191
Bacteriological (Biological) and Toxin Weapons Convention, 65. *See also* Biological Weapons Convention.
Baghdad, 17, 21, 35, 39, 45, 68, 70, 73, 81, 86, 89, 90, 91, 92, 99, 101, 107, 119, 129, 137, 141
Nuclear Research Center, 16, 22, 29, 30
Bahrain, 103
Bailey, Kathleen C., 141
Ballistic missiles. *See* missiles, ballistic and cruise, Iraq.
Ballistic Missile Defense Office, 180
Bari Harbor, 160n
Basra, 85
Battle damage assessment (BDA), 187
Bihac, Bosnia, 105
Biological warfare agents
  Aflatoxin, 74, 76, 77, 78, 81
  Anthrax, 63, 64, 68, 70, 71, 74, 78, 80
  Botulinum, 63, 64, 67, 68, 70, 71, 74, 76, 77, 78, 80, 81
  Brucellas abortus, 70
  Brucella melitensis, 70
  Clostridium perfringens, 70, 74, 77
  Francisella tularensis, 70
Biological weapons, 1, 2, 4, 7, 19, 63, 64, 65, 66, 68, 152, 155, 159, 160, 161, 162, 163, 165, 166, 172, 173, 177, 178, 181, 183, 184
  Characteristics, Table 1-3, 12-14
  Convention, 6n, 7, 151n, 168n
  Export control, 133-35
  Iraq's biological weapons program, 70-79

Proliferation, 145
Black Hole, 45. *See also* U.S. Central Command, Special Planning Group.
Black market, international, 61, 152
Blister agents, 84. *See also* chemical warfare agents.
Blix, Hans, 28
Bombers, long-range, 159, 201-03
Bosnia, 157
  Serbs, 105
Bottom-Up Review, 165, 203
Botulinum, 63, 64, 67, 68, 70, 71, 74, 76, 77, 78, 80, 81. *See also* biological warfare agents.
Brazil, 104
Brookings Institution, 158
Brucellas abortus, 70. *See also* biological warfare agents.
Brucella melitensis, 70. *See also* biological warfare agents.
Brussels, Belgium, 106
Bull, Gerald, 106
Bulletin of Atomic Scientists, 2
Bush Administration, 200
Bush, George, President, 87, 108
Business Week, 5

Calutrons, 25, 47-48, 57, 153
Canada, 106
Capitol Hill, 185
Carter, Ashton, 187
Casualties, aversion to, U.S., 156, 173
Casualty management, chemical and biological agents, 183-84

Central Command, U.S., 11, 68, 87
  Air Force component (CENTAF), 30
  Commander-in-Chief, 11
  Iraqi Target Study, 21
  Operations order (1991), 15
  Special Planning Group (Black Hole), 16. *See also* Black Hole.
Central Europe, 109, 110
Central Intelligence Agency, 58, 124
Centrifuge program, 41, 45, 147
Chemical proliferation, 145
Chemical warfare agents
  Blister agents, 84
  Mustard gas, 84n, 85, 86, 87, 90, 91, 93, 99, 100
  Phosgene, 99
  Sarin, 63, 84, 85, 86, 87, 90, 91, 99
  Soman, 84
  Tabun, 84n, 85, 86, 87, 90, 91
  VX, 86, 91 ,92
Chemical weapons, 1, 2, 4, 7, 18, 19, 152, 155, 159, 160, 162, 163, 165, 166, 172, 173, 177, 178, 181, 183, 184, 186
  Characteristics, 12-14, Table 1-3
  Export control, 133-35
  Iraq's chemical weapons program, 90-99
Chemical Weapons Convention, 6n, 168n
Cheney, Dick, 87
China, 6, 10, 144, 151n, 179

Clinton Administration, 165-70,
171, 182, 183, 185, 187, 196,
198, 199, 203
Clinton, William, President, 144
Clostridium perfringens, 70, 74,
77. *See also* biological
warfare agents.
CNN, 155
Coalition, Western-Arab, 68, 69,
81, 82, 87, 89, 108, 113
Air attacks 18, 22
Air forces, 88, 103, 109
Air planners, 110
Air tasking order, 22
Air strikes, 57, 64, 111
Arab members, 108
Battle damage assessment, 15
Bombing campaign, 16
Counter-Scud operations, 103,
112-13, 125
Counter-nuclear operations, 26
Forces, 22, 101
Military objectives, 11, 15
Scud combat air patrols, 107,
125-26
Target lists, 22
Target planners, 11, 26, 25, 66,
86, 87
Strike plan, 109
Cold War, 1, 6n, 61, 143, 149,
157, 158, 159, 171, 172, 174,
175, 183, 204
Collateral damage, Western
aversion to, 155
Collins, John M., 160
Combat operations, U.S., 159
Command and control centers,
185

Compaq, 5
Congress, U.S., 19
Congressional Research Service,
19, 160
Coordinating Committee on
Multi-Lateral Export control
(COCOM), 6n
Counterforce assets, U.S., 190-95
Counterforce, operational
requirements, 186-98
Counterforce operations, 155,
185-98, 200, 201
Counterforce targeting, 185-86
Counter-missile operations, 108,
109, 110, 127
Counterproliferation, 4, 61, 149
Doctrine, 168
Initiative, 165-70, 175, 197,
199
Operations, 171
Policy, 171
Program, 169, 171
Cruise missiles, 155, 157, 167,
175, 178, 185, 190, 192, 202
Customs barriers, 151
Cuba, 151n

Daura, 74
Dharan, Saudi Arabia, 103
Decontamination, chemical and
biological agents, 183
Deeply-buried bunkers, 155
Defense Intelligence Agency, 18,
25, 112
Automated intelligence File
(AIF), 15-16
Department of Defense, 131,

161, 166, 167, 171, 180, 181, 182, 183, 202, 203
Sales to Iraq, 131, 137
Department of Energy, 58
Department of State, 58, 131
Deployments, 158, 159-61
Deterrence, 171-75, 200, 201-02
Deutch, John, 203
Dijjla, 48
Dual-purpose, 81
Dual-use, 10
Biological, 72
Biological equipment, 134
Capabilities, 119
Chemicals, 134
Commodities and technologies, 6, 7, 129
Definition of, 129n
Items not constrained, 131
Loopholes, 144
Nuclear-related, 133
Sales to Iraq, 131
Technologies, 6n, 134, 143, 144
Dunn, Lewis, 150

East Germany, 110
Egypt, 69, 99, 104, 151n
Ekeus, Rolf, 73, 118
Electro-magnetic isotope separation (EMIS), 28, 39, 41, 45, 47-48, 137, 140
Electro-magnetic pulse (EMP), 184
Escalation dominance, 172
Europe, central, 134
Export controls, Western, 6, 10, 65, 129-44, 151

Ballistic missiles, 135-36
Biological weapons and materials, 133-35
Chemical weapons and materials, 133-35
Circumvention measures, 6, 142, Table 6-2
Computers, 144
International, 130-36
Missiles, 135-36
Missile parts, European firms, 141
Nuclear, 132-33
Export Forum, 6n

F-22 fighter, 203
F-15E fighter, 194
F-16C/D fighter, 194
F-117, fighter, 89, 192-93
F/A-18 fighter, 194, 203
Faw, 106. See also missile, ballistic and cruise, Iraq.
Fenchel, Connie, 143
Flournoy, Michele, 197
Force dispersal (for survivability), 183
Foreign sources of Iraq's supplies, 138-39, Table 6-1
Forward-deployed forces, U.S., 167
France, 16-17, 41, 78, 80, 132
Integrating GPS, 10
Nuclear reactors supplied to Iraq, 56
Uranium enrichment, 41
Francisella tularensis, 70. See also biological warfare agents.
FROG-7, 111, 125

Future Year's Defense Plan, 181

Garf al Naddaf (Location B), 33
Gartoff, Raymond, 158
Gas centrifuges, 28, 41, 137
Gaseous diffusion, 28
Gates, Robert, 124
Geological Survey Institute, 35
Germany, 104, 140, 160n, 174
  Customs, 137
  Integrating GPS, 10
Global Positioning System
  (GPS), 10, 179, 190, 191
Great Britain, 116, 137
"Great Scud Hunt," 107, 108,
  127
Gulf States, 111
Gulf War. *See* Persian Gulf War
Gulf War Airpower Survey, 19

Halabjah, 84
Hardened bunkers, 155
Hassard, John, 150
Hatteen, 48, 49
HAVE NAP missile, AGM-142,
  194n
Hawley, Richard E., General,
  187
Hibbs, Mark, 21
High Altitude Research Project
  (HARP), 106
Highly Enriched Uranium
  (HEU), 29n, 56, 61, 132
Horner, Charles, Lieutenant
  General, 30, 69
Hussein, Saddam, 1, 2, 15, 60,
  64, 68, 73, 80, 83, 108, 113,
  131, 161, 174

India, 132, 149, 151n, 158
Individual protection, against
  biological and chemical
  warfare agents, 184
Intelligence, 21, 82, 89, 103,
  109, 125-27, 153, 200
  Pre-Gulf War, 61, 125, 152
  Signals, 126
  United States, 22, 28, 45, 68,
    86-87, 89, 113, 128, 176n,
    177
  Western, 25, 26, 57, 80
  WMD-related, 66, 88, 152,
    153, 166, 187-89
Interests, U.S., 157n, 172, 173
International Atomic Energy
  Agency (IAEA), 15, 18, 24,
  27, 28, 29, 30, 33, 39, 45, 48,
  49, 53, 56, 58, 60, 132, 153,
  155
  Inspection teams, 26
  On-site inspection reports, 53
  On-site inspections, 34-35
Intervention, U.S., 157
Iran, 10, 100, 126, 144, 151n,
  158, 166
Iran-Iraq War, 83, 84, 85, 99-
  100, 105
Iraq, 10, 166, 174
  Aqba bin Nafi State
    Establishment, 140
  Atomic Energy Commission
    (IAEC), 28, 137
  Ballistic missile program, 53,
    114-25
  Biological warfare program,
    63-82, 107, 155
  Calutrons, 25

Centers of gravity, 15, 66
Chemical weapons program,
    83-102, 107, 155
"Crash program," nuclear
    weapons, 21-22, 56-57
Deception measures, 26
Foreign procurement
    programs, 137-43
Geological Survey Institute, 35
Implosion gun assembly, 48,
    49
Ministry of Military Industry
    and Industrialization, 28, 29,
    73
Nuclear facilities, identified in
    Gulf War, 24, Table 2-1
Nuclear weapons program, 18,
    21, 22, 24, 25, 26, 27-57,
    153, 155
Petrochemical Project 3
    (PC 3), 27-34
Republican Guard, 15
Residual missile force, 114
State Electric Establishment,
    140
State Enterprise for Pesticide
    Production, 85, 91
State Establishment for Heavy
    Engineering Equipment, 48
Uranium enrichment
    programs, 17, 27, 35-41, 153
Weapons of mass destruction,
    1-11, 15, 19, 156
Western export controls,
    deception of, 129-44
Islam, 169

Israel, 16-17, 64, 103, 107, 108,
    110, 111, 125, 126, 150, 151n,
    179
Intelligence on Iraq's nuclear
    weapons, 28
Integrating GPS, 10
Nuclear weapons, 95
Italy, 104, 160n

Jaffar, Jaffar Dhia, 17
Japan, 53
Asahi ion-exchange process,
    45
Joint Air-to-Surface Standoff
    Missile (JASSM), 194n
Joint Direct Attack Munition
    (JDAM), 190, 194
Joint Standoff Weapon (JSOW),
    AGM-154, 194n
Joint Strike Fighter, 202
Jordan, 25, 73, 108

Kahamisyah, 93
Kamel, Hassan, Lieutenant
    General, 73
Kamel, Saddam, Colonel, 73n
Kay, David A., 26
Kirkuk, 89
Knesset, 179
Korean Peninsula, 168
Kurds, 84
Kuwait, 15, 21, 56, 84, 86, 88,
    113, 125, 131, 174, 199

Laith, 125. *See also* missiles,
    ballistic and cruise, Iraq.
Laos, 151n
Laser-guided weapons, 190, 194

Latifiyah, 80
Lawrence Livermore National
    Laboratory, 141
Leadership, U.S., 157
Leventhal, Paul, 58
Libya, 126, 144, 150, 151n, 199
Limited war aims, U.S., 156,
    161
Location B, 33
Long-range strike force, 199-200
Los Alamos National
    Laboratory, 53
Losses in war, U.S., 103

Middle East, 83, 103
Military forces, U.S., 66, 157,
    158, 160, 166, 175, 179, 181,
    182, 184, 194, 202, 203
Deployment, 157
Doctrine, 130
Intervention, 159
Planners, 81
Posture, 156, 163
Power, 172
Strategy Assessment, 147-63
Strategy, options, 165-204
Missiles, ballistic and cruise,
    Iraq, 1, 4, 7, 10, 103-27, 145,
    152, 154, 163, 167, 176, 178,
    179, 181, 186, 188, 202
Ballistic,
    Ababil-100, 124
    Al Abbas, 105
    Al Abid, 105
    Al Fahd 300, 105
    Al Fahd 500, 105
    Al Hijarah, 105

Al Hussein, 90, 104, 105,
    124, 125
Badr-2000, 104
Laith, 125
Scud-B, 26, 63, 103, 104,
    105, 109, 113, 115, 126,
    154, 197
Cruise,
    Export control, 135-36
    Faw, 106
    Silkworm, 106
Missile defenses, U.S., 180
Missile proliferation, 145
Missiles, surface-to-air (SA-2),
    105
Missile Technology Control
    Regime (MTCR), 6n, 135-36
Molecular Laser Isotope
    Separation (M-LIS), 29
Mosul, 39, 124
Muckerman, Joseph E., 173
Muhammadiyat, 92
Munitions List, U.S., 131
Muslim world, 169
Mustard gas, 84n, 85, 86, 87, 90,
    91, 93, 99, 100. See also
    chemical warfare agents.
Myanmar (Burma), 10, 151n

Nassar Works, 41
National Academy of Sciences,
    167
National purpose, U.S., 1
National Security Council, 148,
    197
National security, U.S., 163, 172,
    204
Navy, U.S., 180, 191, 202

Nerlich, Uwe, 174
Nerve agents, 84, 91, 93, 100
New Forum, 144
Non-deterrable adversaries, 173
Nonproliferation, 135, 149, 168, 169
North Atlantic Treaty Organization (NATO), 169
North Korea, 6, 104, 126, 144, 151n, 158, 166, 168, 197
Nuclear Control Institute, 58
Nuclear Deterrence in a Regional Context, 161
Nuclear Non-Proliferation Treaty (NPT), 6n, 27, 132, 151, 168n
Nuclear power plants, 179
Nuclear Suppliers Group, 132, 133
Nuclear weapons, 1, 2, 4, 19, 149, 150, 152, 153, 158, 161, 162, 163, 165, 166
Characteristics, 12-14, Table 1-3
Deterrence, 171-75
Iraq, 21, 27, 33, 53
Nuclear export control, 132-33
Proliferation of, 5, 145, 177
Superiority, U.S., 172
United States, 64, 174

Oak Ridge Plant, 25
Operational Planners, U.S., 19
Operation Desert Shield, 22
Operation Desert Storm, 25, 26, 39, 101, 113, 147, 148, 149, 153, 154, 155, 176n, 188, 192, 194n, 199
Organized crime syndicates, 152

Osirak reactor, 17

Pakistan, 10, 158
Passive defenses, 182-85, 200
Patriot anti-missile missile, 128, 180
Payne, Keith, 174
Pentagon, 21, 178, 181, 203, 204
Perry, William, J., 169, 170, 180, 181
Petrochemical Project 3 (PC 3), 27-34
Pfeiffer, Eckland, 5
Phosgene, 99. See also chemical warfare agents.
Plutonium, 17
Ports and airfields, U.S. dependence on for access, 156, 159
Powell, Colin, General, 87
Power projection capabilities, U.S., 4, 82, 156, 159, 165, 173, 176, 183, 199
Powers, John R., 173
Preemption, 171, 172, 196-99
Preemption paradox, 197-98
Project Babylon, 106, 115
Proliferation, weapons of mass destruction, 1-11, 145

Qayyara, 89

Rabiyah Manufacturing Plant, 47
RAND Corporation, 158, 161
Rashdiah Engineering Center, 41
Rathmell, Andrew, 101
Reagan Administration, 148

Regional WMD proliferation, four conclusions, 147-56
Riyadh, 64, 103
Role in the world, U.S., 1, 2
Russia, 6, 10, 56, 143, 151n

Saddam Works, 41
Salladine, 48
Salman Pak, 68, 70, 71, 88
Samarra, 87, 88, 91
Sandurji, K., General, 149
Satellites, 1, 4, 10, 24, 153
Sarin, 63, 84, 85, 86, 87, 90, 91, 99. *See also* chemical warfare agents.
Saudi Arabia, 64, 103, 107, 110, 111, 125, 161, 179, 198
Schwartzkopf, Norman, General, 11, 15, 30, 87
Scud, 26, 63, 103, 104, 109, 113, 115, 126, 154, 197
Area search boxes, 107
Biological agent-filled warheads, 78
Decoys, 110, 115
Mobile launchers, 108, 109, 111, 154, 187
Operations, 110, 112, 189
Patrols, 111-12, 127
Targets, 16
Warheads, 64, 74. *See also* missiles, ballistic and cruise, Iraq.
Security interests, U.S., 1, 2
Seoul, South Korea, 161
Silkworm, 106. *See also* missiles, ballistic and cruise, Iraq.

Somalia, 157
Soman, 84. *See also* chemical warfare agents.
South Africa, 151n
Southwest Asia, 150
Soviet Union, 69, 103-04, 109, 110, 117, 148, 171, 172
Soviet Union, former, 61, 126, 144, 152
Special operations forces, 111
Standoff Land-Attack Missile (SLAM), AGM-84E, 194n
Strategic defense, U.S., 167
Strike fighters, U.S., 193
Supergun, 106, 115. *See also* Project Babylon.
Surface-to-Air Missiles, 185
Switzerland, 137
Syria, 39, 126, 150, 151n

Tabun, 84n, 85, 86, 87, 90, 91. *See also* chemical warfare agents.
Taiwan, 151n, 179
Taji, 68, 125,
Tarmiya, 39, 45, 48, 140
Teischer, Howard, 148
Tel Aviv, 64, 103
Terrorism, 166
Theater defense, 175-84
Theater High Altitude Area Defense (THAAD), 180-81
Theater missile defenses, 176, 178-81
Theater of operations, U.S. extending its own, 198
Toffler, Alvin and Heidi, 2-5.

Tomahawk Land Attack Missile
(TLAM), 47, 193, 195n, 203
Transporter-erector launchers
(TELs), 109, 110, 154
Turkey, 89

Ubeidi, Amer Rashid, General,
73
Ukraine, 6
United Nations, 1, 30, 169
Baghdad Monitoring and
Verification Center, 74
Food and Agriculture
Organization, 68
Security Council, 73, 118, 143,
168n
Security Council Resolution
687, 18, 26, 27, 114, 115
Special Commission on Iraq
(UNSCOM), 15, 18-19, 64,
66, 69, 70-72, 73, 74, 78, 80,
81, 91, 92-93, 114, 115, 116,
117, 118, 119, 124, 125,
141, 153, 155
United States,
Air Forces in Europe, 187
Bombing raid against Libya
(1986), 198
Customs, 137, 143
General Accounting Office, 78
National purpose, 1
University of London, 150
Unmanned Aerial Vehicles, 10
Uranium enrichment, 17-18,
41-48
Atomic Vapor Laser-Isotope
Separation (AVLIS), 29
Chemical processes, 28

Electromagnetic Isotope
Separation, 28
Gas Centrifuge, 28
Gaseous diffusion, 28
Molecular Laser Isotope
Separation (M-LIS), 29
Uranium hexaflouride, 41
Uranium oxide, 39
Uranium tetrachloride, 39, 48

VX, 86, 91 ,92. See also
chemical warfare agents.

Wallerstein, Mitchel, 171
War and Anti-War, 2
Warsaw Pact, 110
Watman, Kenneth, 161, 162
Weaponization,
Biological agents, 73
Nuclear weapons, 48-49
Weapons of mass destruction.
See nuclear, biological, and
chemical weapons and ballistic
and cruise missile programs.
West Germany, 86
Western-Arab Coalition, 108
Western Europe, 140
Wilkening, Dean, 161, 162
Woolsey, R. James, 124, 179
World Health Organization, 63

Yellowcake, 39
Yemen, 99

Zangger Committee, 132
Zelikow, Philip, 197